Circuit Problems and Solutions

Volume 2: Network Theorems

Circuit Problems and Solutions

Volume 2: Network Theorems

GERARD LIPPIN

HAYDEN BOOK COMPANY, INC.
Rochelle Park, New Jersey

To

Brenda, Jay, and Gary

 Library of Congress Catalog Card Number 66-29784.

Printed in the United States of America.

2	3	4	5	6	7	8	9	PRINTING
77	78	79	80	81	82	83	84	YEAR

Preface

CIRCUIT PROBLEMS AND SOLUTIONS has been written for students studying electrical circuits at the technical institute and junior college level. Since technicians must be able to understand theory and be able to apply it to actual circuits, the objective of this volume is to give a brief review of the principal network theorems, followed by step-by-step solutions to network problems. Therefore, the student can use this book in conjunction with his text for a better understanding of a-c and d-c theory. The problems are presented in a manner that also permits the use of this book without an instructor.

Gerard Lippin

Contents

The Law of Proportionality

VOLTAGE—In a series branch the voltage across any resistive component is found by multiplying the source voltage by the resistive component whose voltage you are trying to find and dividing by the total resistance of the branch (see Fig. A.1):

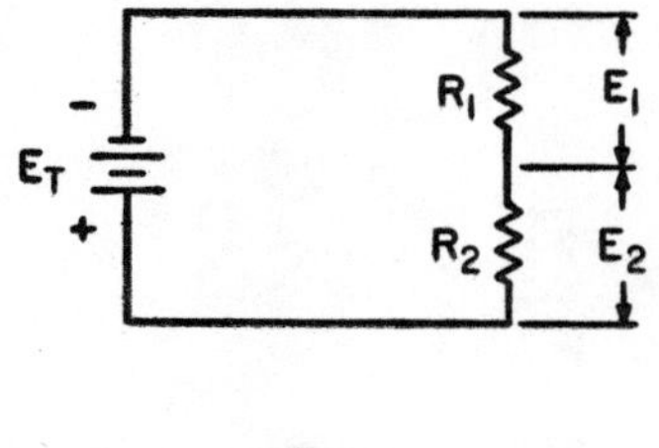

Fig. A.1.

$$E_1 = E_T \frac{R_1}{R_1 + R_2} \qquad E_2 = E_T \frac{R_2}{R_1 + R_2}$$

CURRENT—In a parallel circuit the current flowing through a resistor is found by multiplying the total current by the inverse ratio of the resistive component whose current you are trying to find and dividing by the sum of the resistances in parallel (see Fig. A.2):

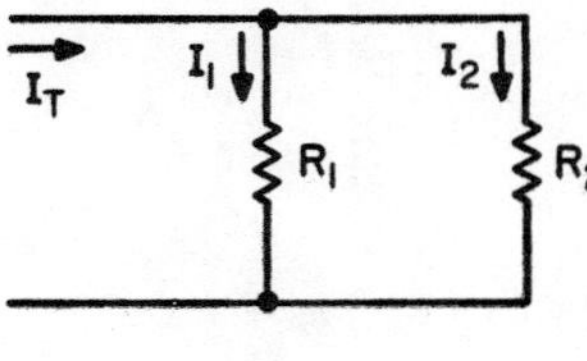

Fig. A.2.

$$I_1 = I_T \frac{R_2}{R_1 + R_2} \qquad I_2 = I_T \frac{R_1}{R_1 + R_2}$$

1

Delta to Wye and Wye to Delta Transformation

Figure 1.1 is called a delta configuration because the circuit is shaped like the Greek letter delta (Δ).

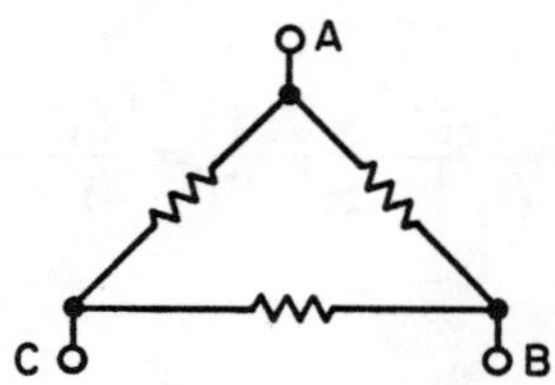

Fig. 1.1.

Figure 1.2 is called a wye ("Y") configuration because schematically it forms a "Y."

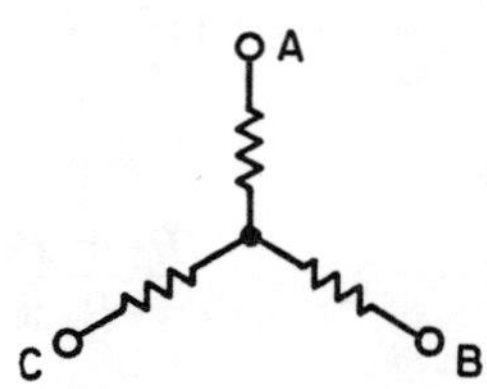

Fig. 1.2.

It is often necessary to convert a circuit from a delta (Δ) configuration to a wye (Y) configuration, or vice-versa, in order to simplify a circuit. Figure 1.3 shows the basic diagram for equivalent delta and wye terminals.

The delta and wye configurations are sometimes referred to as Pi and Tee networks. The formulas developed below are applicable as long as the correct transformations are used.

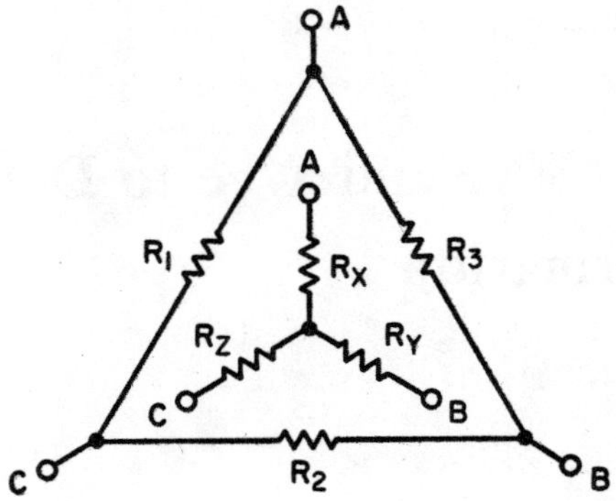

Fig. 1.3.

Converting from delta to wye

Wye:

$$R_{AB} = R_X + R_Y \tag{1.1}$$

Delta:

$$R_{AB} = \frac{R_3[R_1 + R_2]}{R_1 + R_2 + R_3} = \frac{R_1R_3 + R_2R_3}{R_1 + R_2 + R_3} \tag{1.2}$$

Therefore,

$$R_X + R_Y = \frac{R_1R_3 + R_2R_3}{R_1 + R_2 + R_3} \tag{1.3}$$

Similarly,

$$R_X + R_Z = \frac{R_1R_2 + R_1R_3}{R_1 + R_2 + R_3} \tag{1.4}$$

and

$$R_Y + R_Z = \frac{R_1R_2 + R_2R_3}{R_1 + R_2 + R_3} \tag{1.5}$$

To solve for R_X, subtract Eq. 1.5 from Eq. 1.3:

$$R_X - R_Z = \frac{R_1R_3 - R_1R_2}{R_1 + R_2 + R_3} \tag{1.6}$$

Now add Eq. 1.4 to Eq. 1.6:

$$2R_X = \frac{2R_1R_3}{R_1 + R_2 + R_3} \tag{1.7}$$

Dividing both sides by 2 yields:

$$R_X = \frac{R_1 R_3}{R_1 + R_2 + R_3} = \frac{R_{ADJ}}{R_T} \tag{1.8}$$

in which:

R_{ADJ} = multiplication of adjacent resistors
R_T = total resistance, $R_1 + R_2 + R_3$

Similarly,

$$R_Y = \frac{R_{ADJ}}{R_T} = \frac{R_2 R_3}{R_1 + R_2 + R_3} \tag{1.9}$$

and

$$R_Z = \frac{R_{ADJ}}{R_T} = \frac{R_1 R_2}{R_1 + R_2 + R_3} \tag{1.10}$$

Converting from wye to delta

Short out terminals A to B in Fig. 1.3.
Delta:

$$R_{AC} = \frac{R_1 R_2}{R_1 + R_2} \tag{1.11}$$

Wye:

$$\begin{aligned} R_{AC} &= \frac{R_X R_Y}{R_X + R_Y} + R_Z \\ &= \frac{R_X R_Y + R_X R_Z + R_Y R_Z}{R_X + R_Y} \\ &= \frac{R_{SOP}}{R_X + R_Y} \end{aligned} \tag{1.12}$$

in which:

R_{SOP} = sum of products
$= R_X R_Y + R_X R_Z + R_Y R_Z$

Therefore,

$$\frac{R_1 R_2}{R_1 + R_2} = \frac{R_{SOP}}{R_X + R_Y} \tag{1.13}$$

Similarly,

$$\frac{R_1R_3}{R_1 + R_3} = \frac{R_{SOP}}{R_Y + R_Z} \tag{1.14}$$

and

$$\frac{R_2R_3}{R_2 + R_3} = \frac{R_{SOP}}{R_X + R_Z} \tag{1.15}$$

To solve for R_1 invert Eqs. 1.13, 1.14, and 1.15. Thus,

$$\frac{1}{R_1} + \frac{1}{R_2} = \frac{R_X}{R_{SOP}} + \frac{R_Y}{R_{SOP}} \tag{1.16}$$

$$\frac{1}{R_1} + \frac{1}{R_3} = \frac{R_Y}{R_{SOP}} + \frac{R_Z}{R_{SOP}} \tag{1.17}$$

$$\frac{1}{R_2} + \frac{1}{R_3} = \frac{R_X}{R_{SOP}} + \frac{R_Z}{R_{SOP}} \tag{1.18}$$

Subtract Eq. 1.18 from Eq. 1.17, which yields:

$$\frac{1}{R_1} - \frac{1}{R_2} = \frac{R_Y}{R_{SOP}} - \frac{R_X}{R_{SOP}} \tag{1.19}$$

Add Eq. 1.19 to Eq. 1.16, which yields:

$$\frac{2}{R_1} = \frac{2R_Y}{R_{SOP}} \tag{1.20}$$

Solving for R_1 yields:

$$R_1 = \frac{R_{SOP}}{R_Y} = \frac{R_{SOP}}{R_{OPP}} \tag{1.21}$$

in which:

R_{SOP} = sum of products ($R_XR_Y + R_XR_Z + R_YR_Z$)
R_{OPP} = opposite resistor (R_Y in this case)

Similarly,

$$R_2 = \frac{R_{SOP}}{R_{OPP}} = \frac{R_XR_Y + R_XR_Z + R_YR_Z}{R_X} \tag{1.22}$$

and

$$R_3 = \frac{R_{SOP}}{R_{OPP}} = \frac{R_XR_Y + \mathrm{R_XR_Z} + R_YR_Z}{R_Z} \tag{1.23}$$

Problems

Note: It may be possible to solve some of the following problems by converting them either from the delta to the wye configuration or from the wye to the delta configuration. However, we will illustrate only one method in each problem.

1.1 Find the equivalent wye configuration of Fig. 1.4.

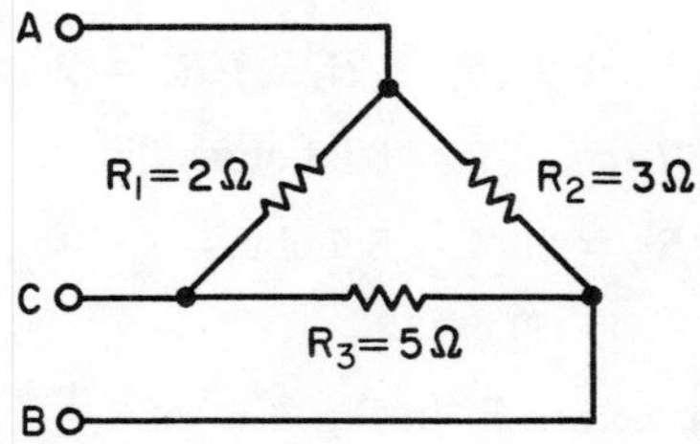

Fig. 1.4.

Answer

See Fig. 1.4a.

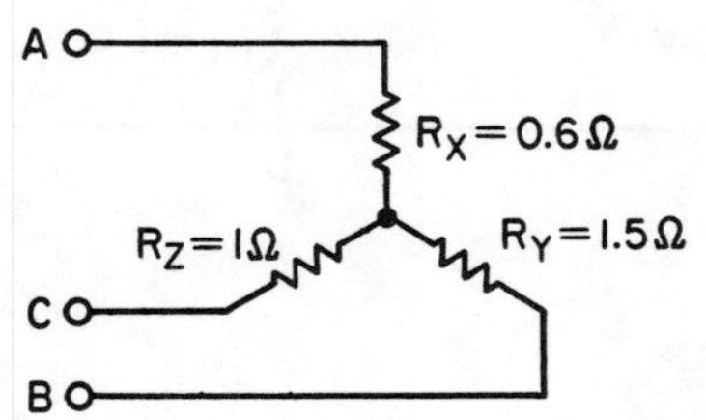

Fig. 1.4a.

$$R_X = R_{ADJ}/R_T = R_1R_2/(R_1 + R_2 + R_3) = (2 \times 3)/(2 + 3 + 5) = 0.6 \text{ ohm}$$

$$R_Y = R_{ADJ}/R_T = (3 \times 5)/10 = 1.5 \text{ ohms}$$

$$R_Z = R_{ADJ}/R_T = (2 \times 5)/10 = 1 \text{ ohm}$$

1.2 Find the equivalent delta (Pi) configuration of Fig. 1.5.

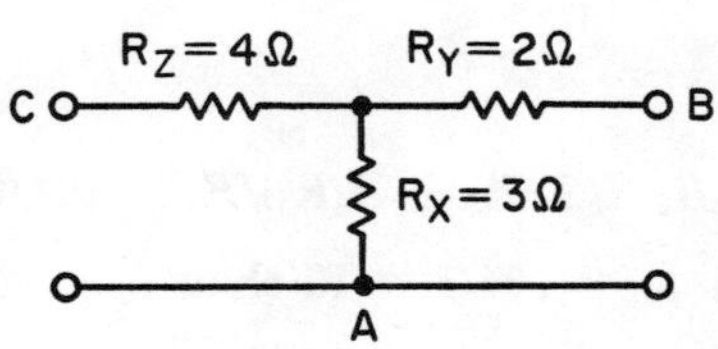

Fig. 1.5.

Answer

See Fig. 1.5a.

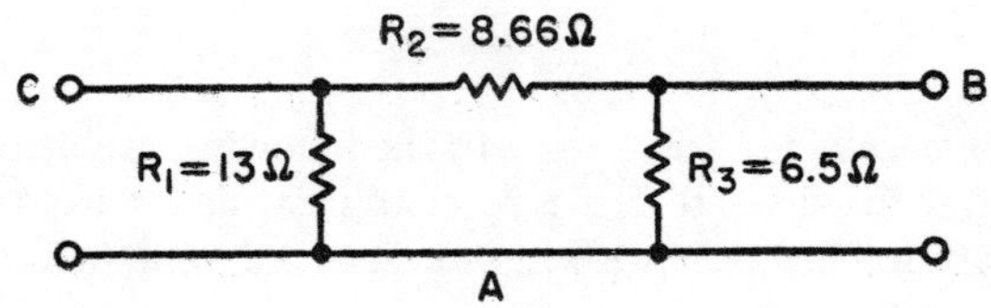

Fig. 1.5a.

$R_1 = R_{SOP}/R_{OPP} = (R_XR_Y + R_XR_Z + R_YR_Z)/R_Y = 26/2 = 13$ ohms

$R_2 = R_{SOP}/R_{OPP} = 26/R_X = 26/3 = 8.66$ ohms

$R_3 = R_{SOP}/R_{OPP} = 26/R_Z = 26/4 = 6.5$ ohms

1.3 Find the voltage drop across R_1 in Fig. 1.6.

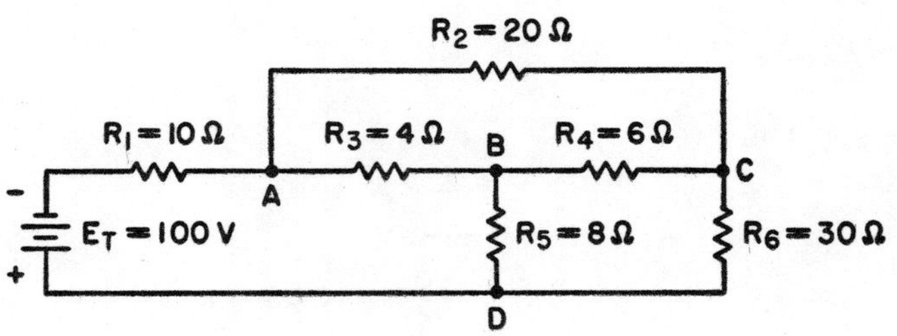

Fig. 1.6.

Answer

Convert the wye consisting of R_3, R_4, and R_5 to a delta (Fig. 1.6a).

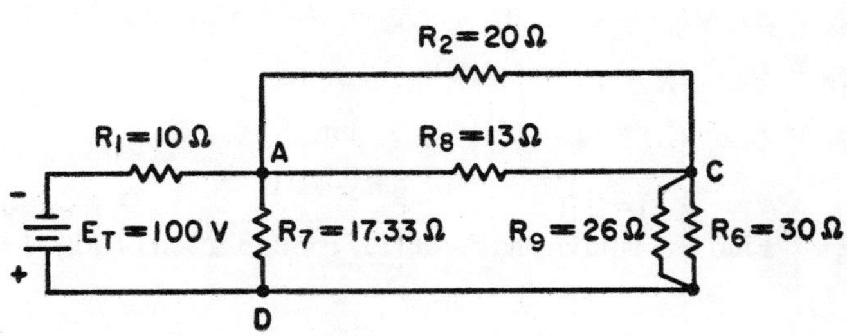

Fig. 1.6a.

$R_7 = R_{SOP}/R_{OPP} = (R_3R_4 + R_3R_5 + R_4R_5)/R_4 = 104/6 = 17.33$ ohms

$R_8 = R_{SOP}/R_{OPP} = 104/R_5 = 104/8 = 13$ ohms

$R_9 = R_{SOP}/R_{OPP} = 104/R_3 = 104/4 = 26$ ohms

R_2 and R_8 are in parallel and form R_{10}. R_6 and R_9 are in parallel and form R_{11} (Fig. 1.6b).

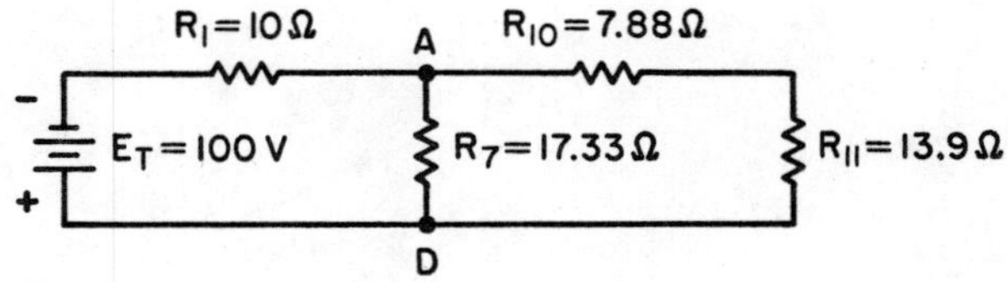

Fig. 1.6b.

$$R_{10} = R_8R_2/(R_8 + R_2) = (13 \times 20)/(13 + 20) = 7.88 \text{ ohms}$$

$$R_{11} = R_9R_6/(R_9 + R_6) = (26 \times 30)/(26 + 30) = 13.9 \text{ ohms}$$

R_{10} and R_{11} are in series and are in parallel with R_7 to form R_{12} (Fig. 1.6c).

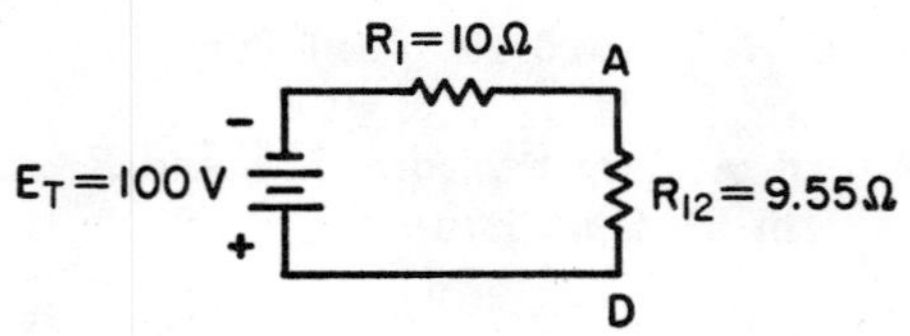

Fig. 1.6c.

$$R_{12} = \frac{(R_{10} + R_{11})R_7}{R_{10} + R_{11} + R_7} = \frac{(7.88 + 13.9)(17.33)}{7.88 + 13.9 + 17.33} = 9.65 \text{ ohms}$$

Using the law of proportionality,

$$E_{R1} = E_T \frac{R_1}{R_1 + R_{12}} = 100 \times \frac{10}{10 + 9.55} = 51.2 \text{ volts}$$

1.4 Find the current flowing through R_8 in Fig. 1.7.

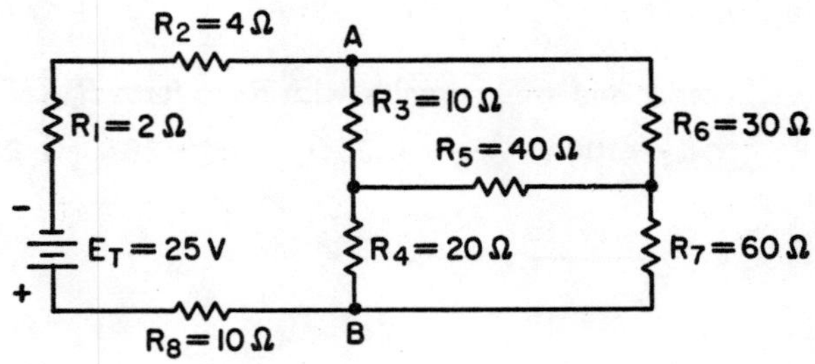

Fig. 1.7.

Answer

Convert the wye consisting of R_3, R_4, and R_5 to a delta (Fig. 1.7a).

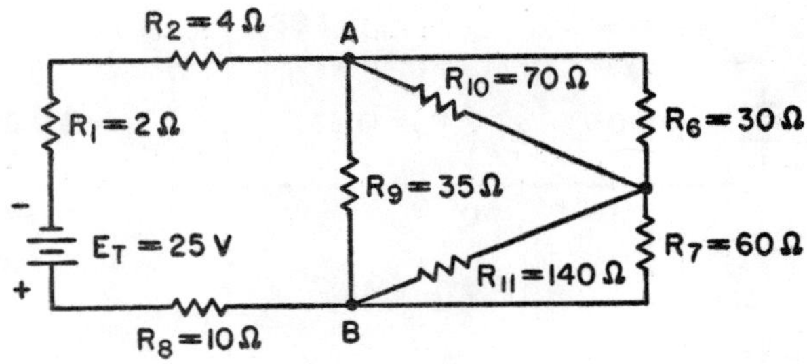

Fig. 1.7a.

$$R_9 = R_{SOP}/R_{OPP} = (R_3R_4 + R_3R_5 + R_4R_5)/R_5 = 1{,}400/40 = 35 \text{ ohms}$$
$$R_{10} = R_{SOP}/R_{OPP} = 1{,}400/R_4 = 1{,}400/20 = 70 \text{ ohms}$$
$$R_{11} = R_{SOP}/R_{OPP} = 1{,}400/R_3 = 1{,}400/10 = 140 \text{ ohms}$$

R_6 and R_{10} are in parallel with each other to form R_{12}, and R_7 and R_{11} are in parallel with each other to form R_{13} (Fig. 1.7b).

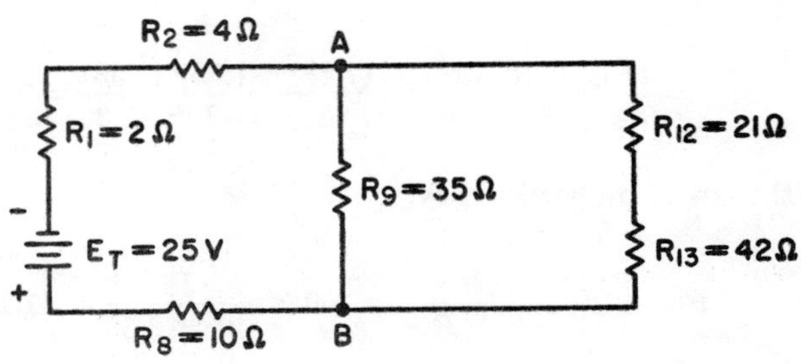

Fig. 1.7b.

$$R_{12} = R_6R_{10}/(R_6 + R_{10}) = (30 \times 70)/(30 + 70) = 21 \text{ ohms}$$
$$R_{13} = R_7R_{11}/(R_7 + R_{11}) = (60 \times 140)/(60 + 140) = 42 \text{ ohms}$$

R_{12} and R_{13} are in series and are in parallel with R_9 to form R_{AB} (Fig. 1.7c).

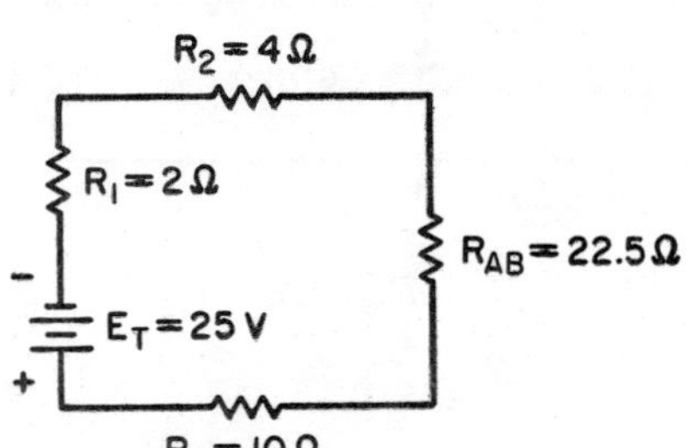

Fig. 1.7c.

$$R_{AB} = \frac{(R_{12} + R_{13})R_9}{R_{12} + R_{13} + R_9} = \frac{(21 + 42)(35)}{21 + 42 + 35} = 22.5 \text{ ohms}$$

$$I_8 = I_T = \frac{E_T}{R_T} = \frac{E_T}{R_1 + R_2 + R_{AB} + R_8} = \frac{25}{(2 + 4 + 22.5 + 10)} = 0.65 \text{ amp}$$

1.5 Find the voltage across R_6 in Fig. 1.8.

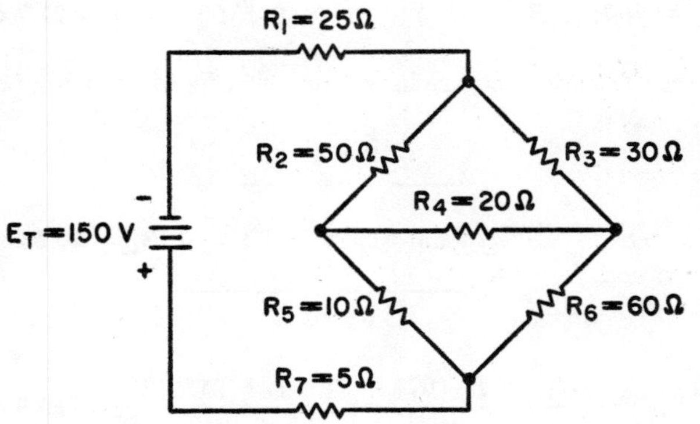

Fig. 1.8.

Answer

Convert the wye consisting of R_2, R_4, and R_5 to a delta (Fig. 1.8a).

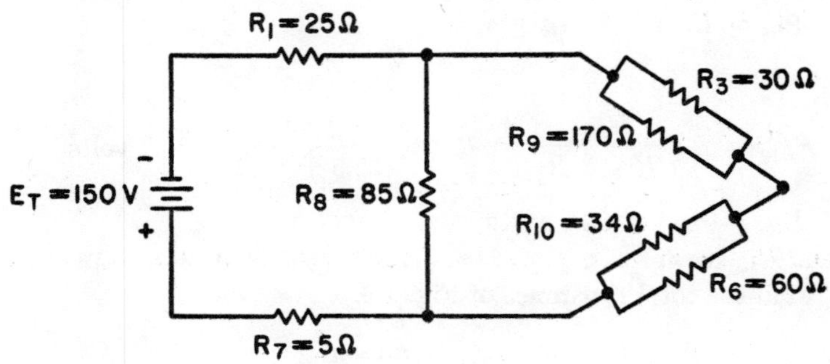

Fig. 1.8a.

$$R_8 = R_{SOP}/R_{OPP} = (R_2R_4 + R_2R_5 + R_4R_5)/R_4 = 1{,}700/20 = 85 \text{ ohms}$$

$$R_9 = R_{SOP}/R_{OPP} = 1{,}700/R_5 = 1{,}700/10 = 170 \text{ ohms}$$

$$R_{10} = R_{SOP}/R_{OPP} = 1{,}700/R_2 = 1{,}700/50 = 34 \text{ ohms}$$

R_3 and R_9 are in parallel with each other to form R_{11}, and R_6 and R_{10} are in parallel with each other to form R_{12} (Fig. 1.8b).

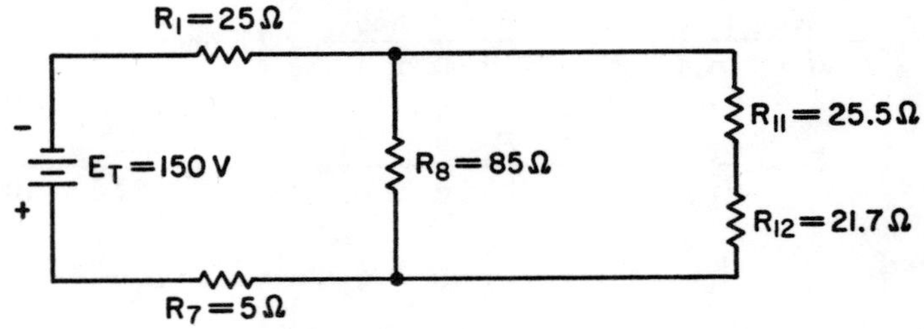

Fig. 1.8b.

$$R_{11} = R_3R_9/(R_3 + R_9) = (30 \times 170)/(30 + 170) = 25.5 \text{ ohms}$$

$$R_{12} = R_6R_{10}/(R_6 + R_{10}) = (60 \times 34)/(60 + 34) = 21.7 \text{ ohms}$$

R_{11} and R_{12} are in series and are in parallel with R_8 to form R_{13} (Fig. 1.8c).

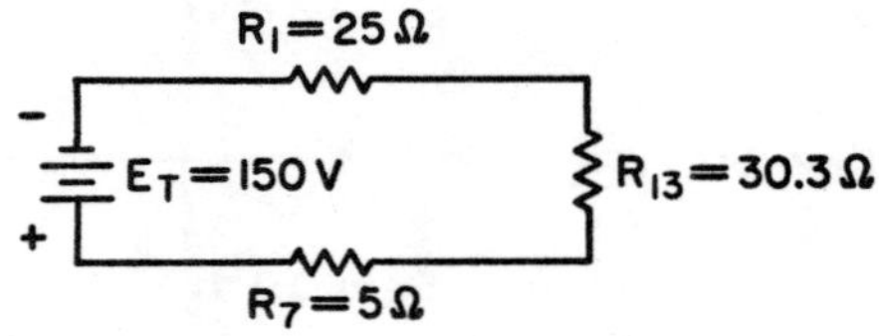

Fig. 1.8c.

$$R_{13} = \frac{(R_{11} + R_{12})R_8}{R_{11} + R_{12} + R_8} = \frac{(25.5 + 21.7)(85)}{25.5 + 21.7 + 85} = 30.3 \text{ ohms}$$

R_1, R_{13}, and R_7 are in series with each other. Using the law of proportionality,

$$E_{R13} = E_T \frac{R_{13}}{R_1 + R_{13} + R_7} = 150 \times \frac{30.3}{25 + 30.3 + 5} = 75.4 \text{ volts}$$

$$E_{R8} = E_{R13} = 75.4 \text{ volts}$$

Again using

$$E_{R12} = E_{R8} \frac{R_{12}}{R_{11} + R_{12}} = 75.4 \times \frac{21.7}{25.5 + 21.7} = 34.7 \text{ volts}$$

$$E_{R6} = E_{R12} = 34.7 \text{ volts}$$

1.6 Find the total resistance of Fig. 1.9.

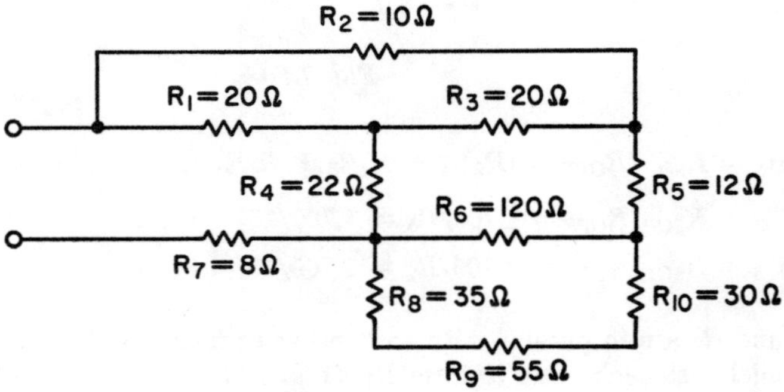

Fig. 1.9.

Answer

Convert the delta consisting of R_1, R_2, and R_3 to a wye (Fig. 1.9a).

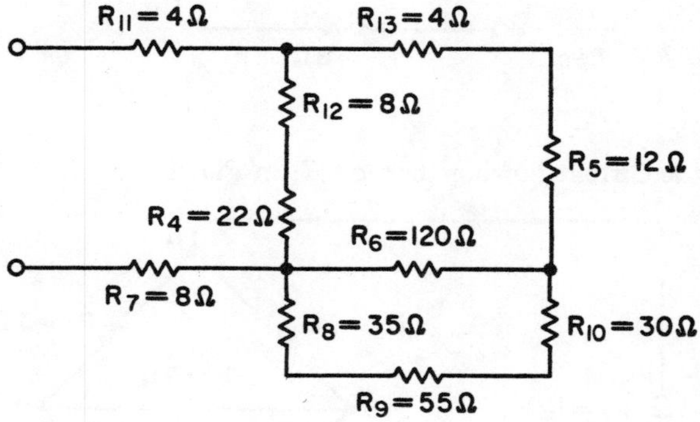

Fig. 1.9a.

$$R_{11} = R_{ADJ}/R_T = R_1R_2/(R_1 + R_2 + R_3) = (20 \times 10)/(20 + 10 + 20) = 4 \text{ ohms}$$

$$R_{12} = R_{ADJ}/R_T = R_1R_3/50 = (20 \times 20)/50 = 8 \text{ ohms}$$

$$R_{13} = R_{ADJ}/R_T = R_2R_3/50 = (10 \times 20)/50 = 4 \text{ ohms}$$

R_4 is in series with R_{12} to form R_{14}, and R_5 is in series with R_{13} to form R_{15}. R_8, R_9, and R_{10} are in series and are in parallel with R_6 to form R_{16} (Fig. 1.9b).

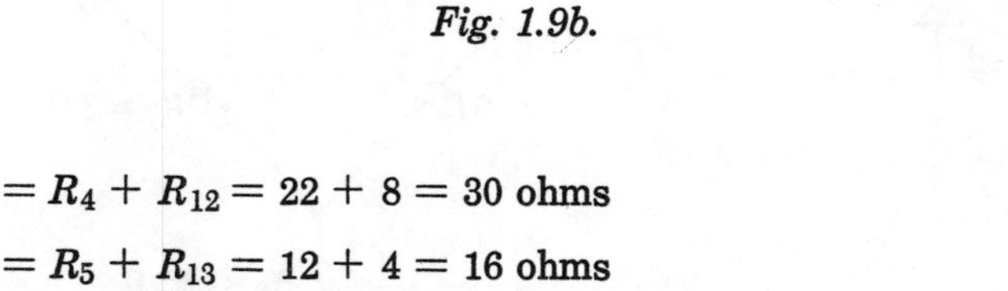

Fig. 1.9b.

$$R_{14} = R_4 + R_{12} = 22 + 8 = 30 \text{ ohms}$$

$$R_{15} = R_5 + R_{13} = 12 + 4 = 16 \text{ ohms}$$

$$R_{16} = \frac{(R_8 + R_9 + R_{10})R_6}{R_8 + R_9 + R_{10} + R_6} = \frac{(35 + 55 + 30)(120)}{35 + 55 + 30 + 120} = 60 \text{ ohms}$$

R_{15} and R_{16} are in series and are in parallel with R_{14} to form R_{17} (Fig. 1.9c).

$$R_{17} = \frac{(R_{15} + R_{16})R_{14}}{R_{15} + R_{16} + R_{14}} = \frac{(16 + 60)(30)}{16 + 60 + 30} = 21.5 \text{ ohms}$$

$$R_T = R_{11} + R_{17} + R_7 = 4 + 21.5 + 8 = 33.5 \text{ ohms}$$

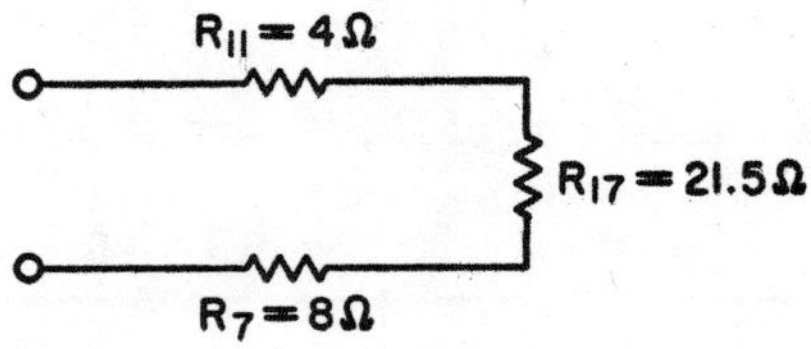

Fig. 1.9c.

1.7 Find the current flowing through R_1 in Fig. 1.10.

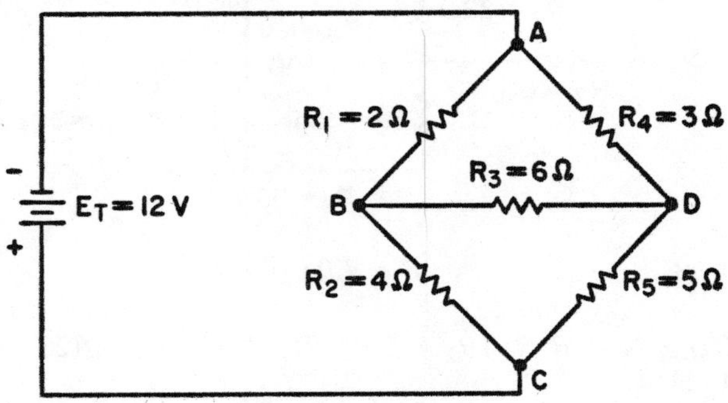

Fig. 1.10.

Answer

Convert the delta consisting of R_2, R_3, and R_5 to a wye (Fig. 1.10a).

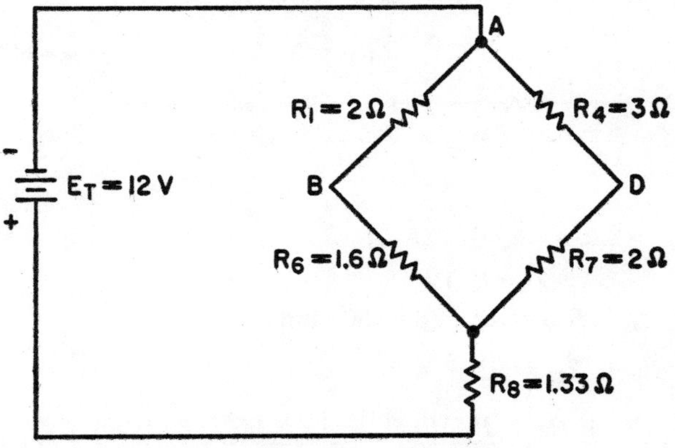

Fig. 1.10a.

$R_6 = R_{ADJ}/R_T = R_2R_3/(R_2 + R_3 + R_5) = (4 \times 6)/(4 + 6 + 5) = 1.6$ ohms

$R_7 = R_{ADJ}/R_T = R_3R_5/15 = (6 \times 5)/15 = 2$ ohms

$R_8 = R_{ADJ}/R_T = R_2R_5/15 = (4 \times 5)/15 = 1.33$ ohms

The series combination of R_1 and R_6 is in parallel with the series combination of R_4 and R_7 to form R_9 (Fig. 1.10b).

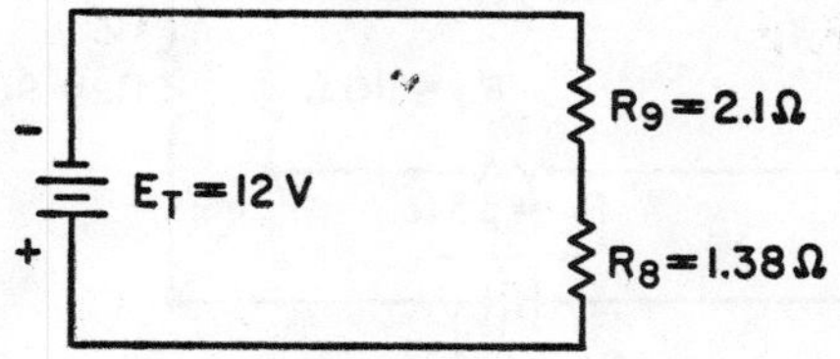

Fig. 1.10b.

$$R_9 = \frac{(R_1 + R_6)(R_4 + R_7)}{R_1 + R_6 + R_4 + R_7} = \frac{(2 + 1.6)(3 + 2)}{(2 + 1.6) + (3 + 2)} = 2.1 \text{ ohms}$$

Using the law of proportionality,

$$E_{R9} = E_T \frac{R_9}{R_8 + R_9} = 12 \times \frac{2.1}{1.33 + 2.1} = 7.3 \text{ volts}$$

The voltage across R_9 is equal to the voltage across the sum of resistors R_1 and R_6 and thus:

$$I_{R1} = E_{R9}/(R_1 + R_6) = 7.3/(2 + 1.6) = 2.03 \text{ amps}$$

1.8 Find the total resistance of the circuit shown in Fig. 1.11.

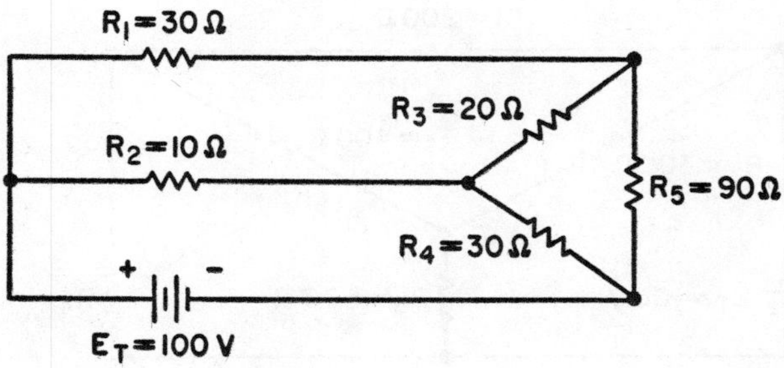

Fig. 1.11.

Answer

Convert the wye consisting of R_2, R_3, and R_4 to a delta (Fig. 1.11a).

$$R_6 = R_{SOP}/R_{OPP} = (R_2R_3 + R_2R_4 + R_3R_4)/R_4 = 1{,}100/30 = 36.7 \text{ ohms}$$

$$R_7 = R_{SOP}/R_{OPP} = 1{,}100/R_2 = 1{,}100/10 = 110 \text{ ohms}$$

$$R_8 = R_{SOP}/R_{OPP} = 1{,}100/R_3 = 1{,}100/20 = 55 \text{ ohms}$$

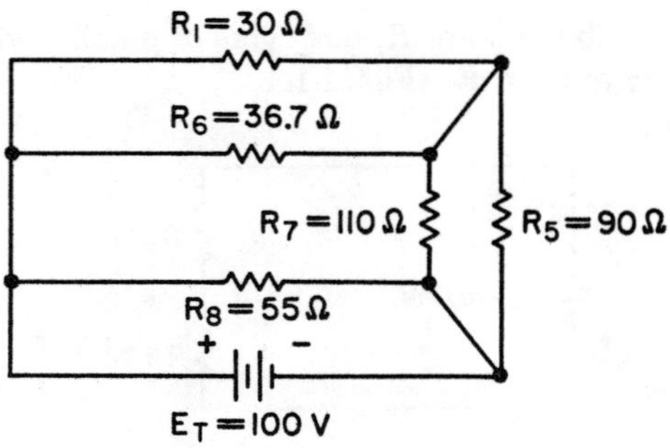

Fig. 1.11a.

The parallel combination of R_1 and R_6 is in series with the parallel combination of R_7 and R_5, which in turn is in parallel with R_8 to form R_T:

$$R_T = \frac{\left[\frac{R_1R_6}{R_1+R_6}+\frac{R_7R_5}{R_7+R_5}\right]R_8}{\frac{R_1R_6}{R_1+R_6}+\frac{R_7R_5}{R_7+R_5}+R_8} = \frac{\left[\frac{30\times 36.7}{30+36.7}+\frac{110\times 90}{110+90}\right]55}{\frac{30\times 36.7}{30+36.7}+\frac{110\times 90}{110+90}+55}$$

$$R_T = \frac{(16.7+49.5)55}{16.7+49.5+55} = \frac{66.3\times 55}{121.2} = 30 \text{ ohms}$$

1.9 Find the current flowing through R_4 in Fig. 1.12.

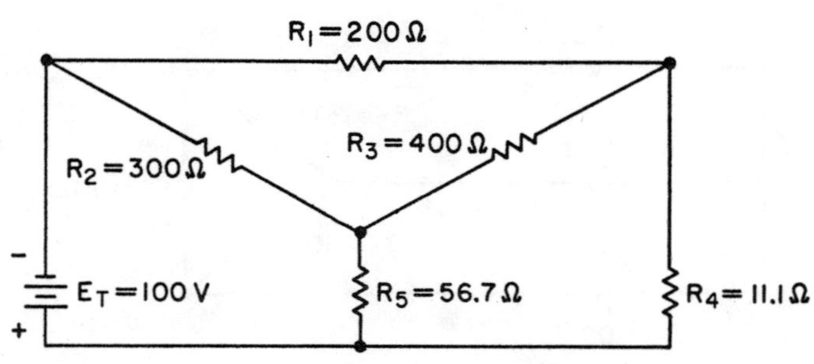

Fig. 1.12.

Answer

Convert the delta consisting of R_1, R_2, and R_3 to a wye (Fig. 1.12a).

$R_6 = R_{ADJ}/R_T = R_1R_2/(R_1 + R_2 + R_3) = (200 \times 300)/(200 + 300 + 400)$
$= 66.7$ ohms

$R_7 = R_{ADJ}/R_T = R_1R_3/900 = (200 \times 400)/900 = 88.9$ ohms

$R_8 = R_{ADJ}/R_T = R_2R_3/900 = (300 \times 400)/900 = 133.3$ ohms

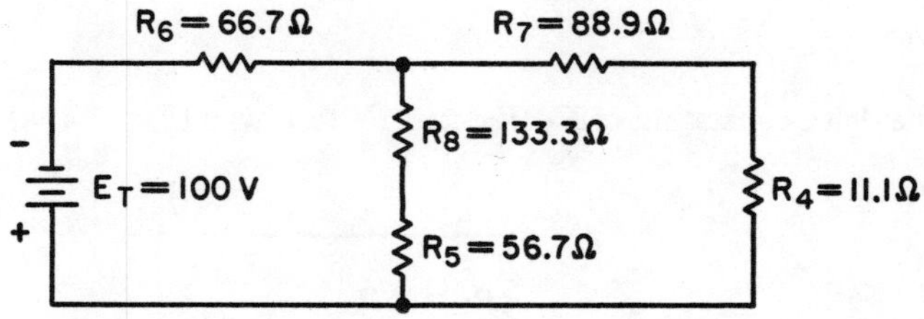

Fig. 1.12a.

The series combination of R_7 and R_4 is in parallel with the series combination of R_8 and R_5 to form R_9 (Fig. 1.12b).

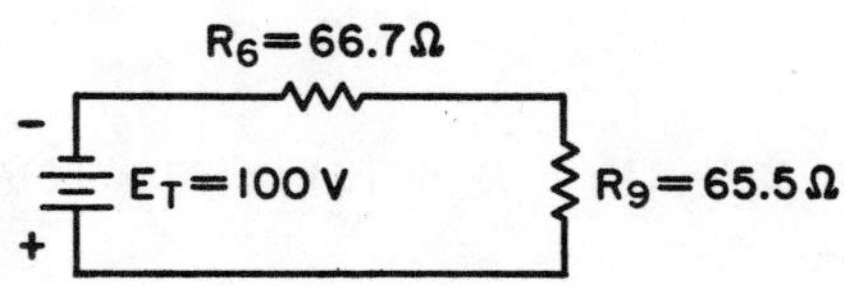

Fig. 1.12b.

$$R_9 = \frac{(R_7 + R_4)(R_8 + R_5)}{R_7 + R_4 + R_8 + R_5} = \frac{(88.9 + 11.1)(133.3 + 56.7)}{88.9 + 11.1 + 133.3 + 56.7} = 65.5 \text{ ohms}$$

Using the law of proportionality,

$$E_{R9} = E_T \frac{R_9}{R_6 + R_9} = 100 \times \frac{65.5}{66.7 + 65.5} = 49.5 \text{ volts}$$

The voltage across R_9 is equal to the voltage across the sum of the resistors R_7 and R_4. Thus:

$$I_{R4} = E_{R9}/(R_7 + R_4) = 49.5/(88.9 + 11.1) = 0.495 \text{ amp}$$

1.10 Find the current flowing through R_6 in Fig. 1.13.

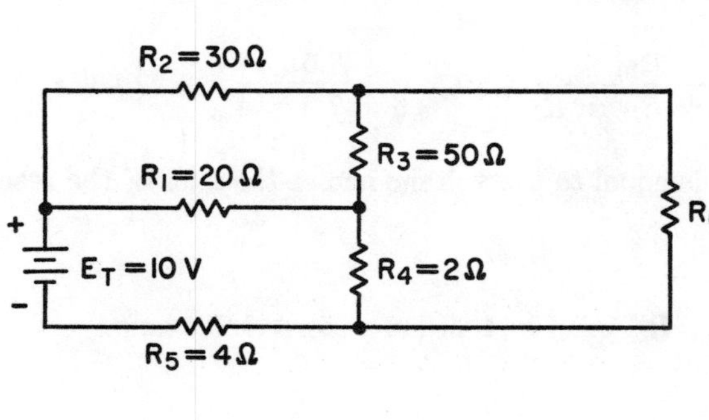

Fig. 1.13.

Answer

Convert the delta consisting of R_1, R_2, and R_3 to a wye (Fig. 1.13a).

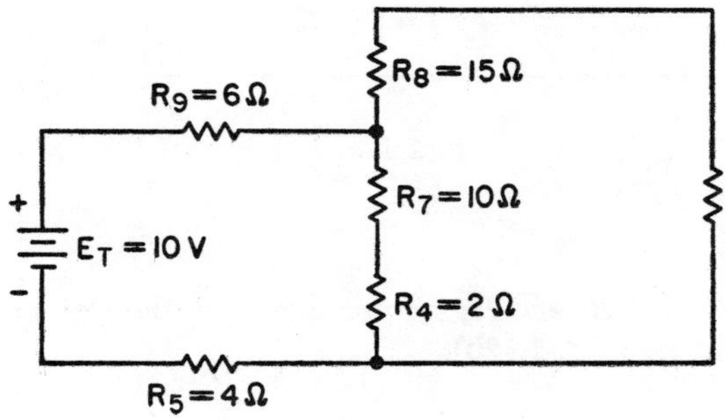

Fig. 1.13a.

$$R_7 = R_{ADJ}/R_T = R_1R_3/(R_1 + R_2 + R_3) = (20 \times 50)/(20 + 30 + 50) = 10 \text{ ohms}$$

$$R_8 = R_{ADJ}/R_T = R_2R_3/100 = (30 \times 50)/100 = 15 \text{ ohms}$$

$$R_9 = R_{ADJ}/R_T = R_1R_2/100 = (20 \times 30)/100 = 6 \text{ ohms}$$

The series combination of R_6 and R_8 is in parallel with the series combination of R_4 and R_7 to form R_{10} (Fig. 1.13b).

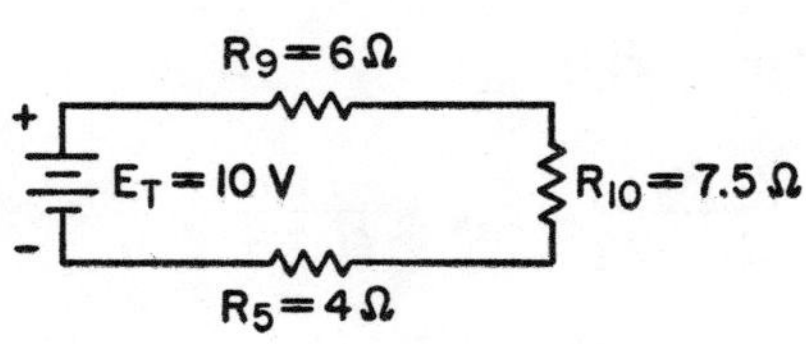

Fig. 1.13b.

$$R_{10} = \frac{(R_8 + R_6)(R_4 + R_7)}{R_8 + R_6 + R_4 + R_7} = \frac{(15 + 5)(2 + 10)}{15 + 5 + 2 + 10} = 7.5 \text{ ohms}$$

Using the law of proportionality,

$$E_{R10} = E_T \frac{R_{10}}{R_9 + R_{10} + R_5} = 10 \times \frac{7.5}{6 + 7.5 + 4} = 4.29 \text{ volts}$$

The voltage across R_{10} is equal to the voltage across the sum of the resistors R_6 and R_8. Thus,

$$I_{R6} = E_{R10}/(R_8 + R_6) = 4.29/(15 + 5) = 0.214 \text{ amp}$$

1.11 Find the current drawn from the battery in Fig. 1.14.

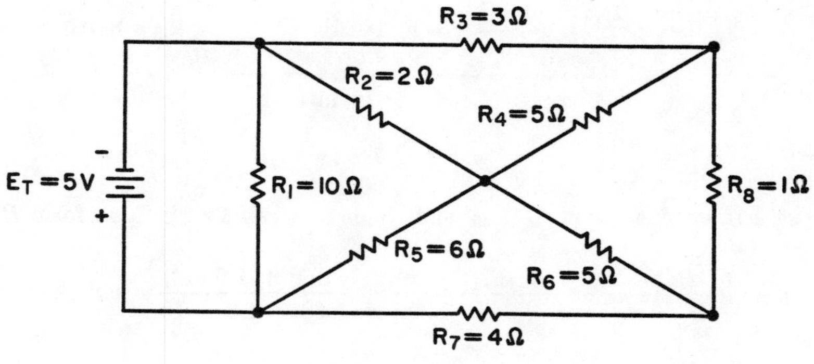

Fig. 1.14.

Answer

Convert the delta consisting of R_2, R_3, and R_4 to a wye and the delta consisting of R_5, R_6, and R_7 to a wye (Fig. 1.14a).

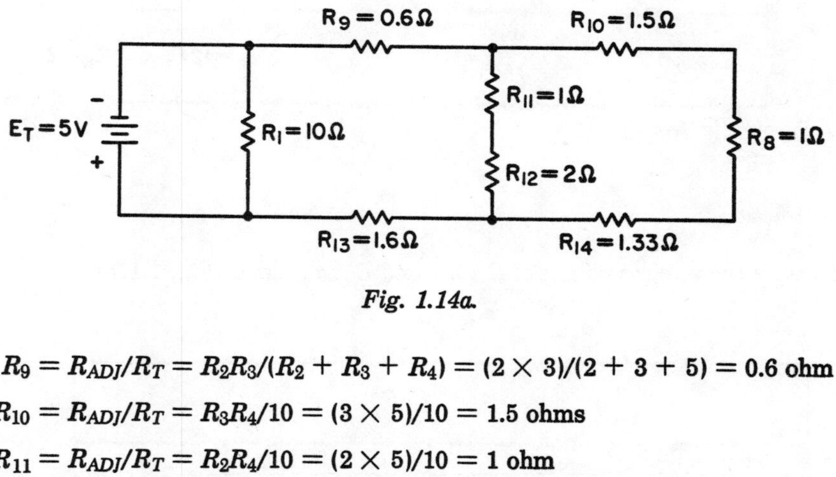

Fig. 1.14a.

$$R_9 = R_{ADJ}/R_T = R_2R_3/(R_2 + R_3 + R_4) = (2 \times 3)/(2 + 3 + 5) = 0.6 \text{ ohm}$$

$$R_{10} = R_{ADJ}/R_T = R_3R_4/10 = (3 \times 5)/10 = 1.5 \text{ ohms}$$

$$R_{11} = R_{ADJ}/R_T = R_2R_4/10 = (2 \times 5)/10 = 1 \text{ ohm}$$

$$R_{12} = R_{ADJ}/R_T = R_5R_6/(R_5 + R_6 + R_7) = (6 \times 5)/(6 + 5 + 4) = 2 \text{ ohms}$$

$$R_{13} = R_{ADJ}/R_T = R_5R_7/15 = (6 \times 4)/15 = 1.6 \text{ ohms}$$

$$R_{14} = R_{ADJ}/R_T = R_6R_7/15 = (5 \times 4)/15 = 1.33 \text{ ohms}$$

The series combination of R_{10}, R_8, and R_{14} is in parallel with the series combination of R_{11} and R_{12} to form R_{15} (Fig. 1.14b).

$$R_{15} = \frac{(R_{10} + R_8 + R_{14})(R_{11} + R_{12})}{R_{10} + R_8 + R_{14} + R_{11} + R_{12}} = \frac{(1.5 + 1 + 1.33)(1 + 2)}{1.5 + 1 + 1.33 + 1 + 2}$$
$$= 1.68 \text{ ohms}$$

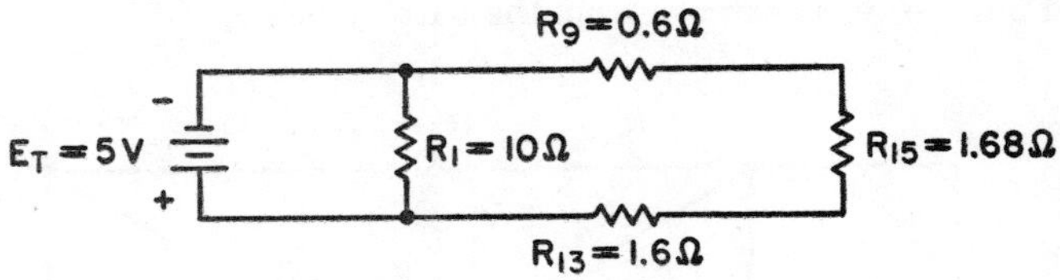

Fig. 1.14b.

The series combination of R_9, R_{15}, and R_{13} is in parallel with R_1 to form R_T:

$$R_T = \frac{(R_9 + R_{15} + R_{13})R_1}{R_9 + R_{15} + R_{13} + R_1} = \frac{(0.6 + 1.68 + 1.6)(10)}{0.6 + 1.68 + 1.6 + 10} = 2.8 \text{ ohms}$$

$$I_T = E_T/R_T = 5/2.8 = 1.8 \text{ amps}$$

1.12 Find the current flowing through R_5 in Fig. 1.15.

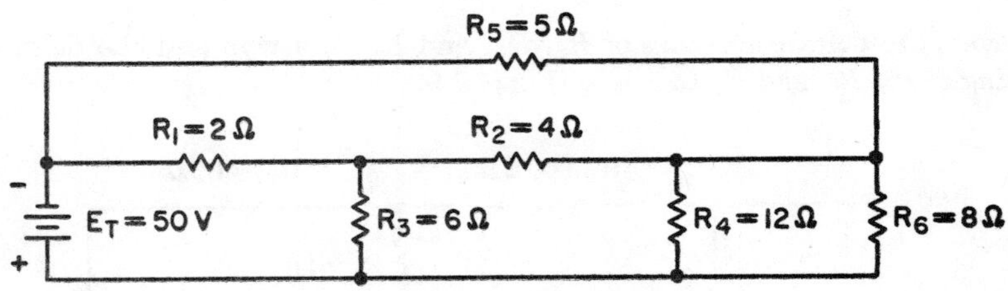

Fig. 1.15.

Answer

Convert the wye consisting of R_1, R_2, and R_3 to a delta (Fig. 1.15a).

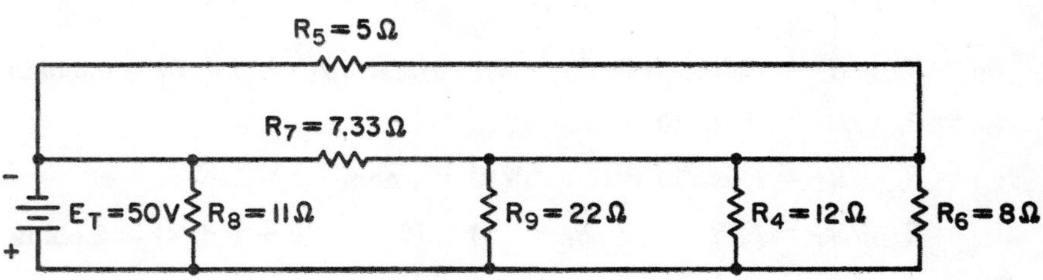

Fig. 1.15a.

$$R_7 = R_{SOP}/R_{OPP} = (R_1R_2 + R_1R_3 + R_2R_3)/R_3 = 44/6 = 7.33 \text{ ohms}$$

$$R_8 = R_{SOP}/R_{OPP} = 44/R_2 = 44/4 = 11 \text{ ohms}$$

$$R_9 = R_{SOP}/R_{OPP} = 44/R_1 = 44/2 = 22 \text{ ohms}$$

R_9, R_4, and R_6 are in parallel with each other to form R_{11}, and R_5 and R_7 are in parallel with each other to form R_{10} (Fig. 1.15b).

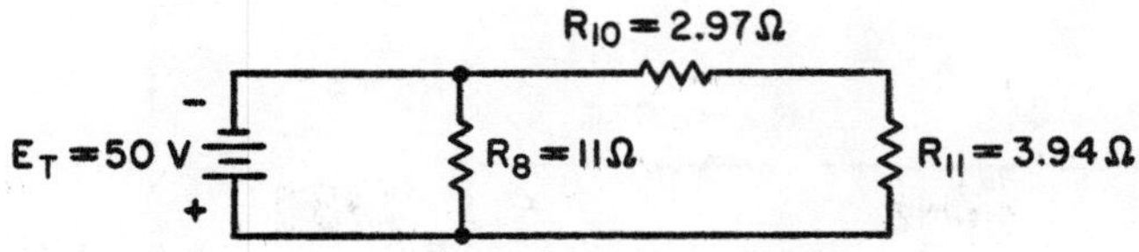

Fig. 1.15b.

$$R_{10} = R_5R_7/(R_5 + R_7) = (5 \times 7.33)/(5 + 7.33) = 2.97 \text{ ohms}$$

$$R_{11} = \frac{1}{\frac{1}{R_9}+\frac{1}{R_4}+\frac{1}{R_6}} = \frac{1}{\frac{1}{22}+\frac{1}{12}+\frac{1}{8}} = 3.94 \text{ ohms}$$

The voltage across the sum of the resistors, R_{10} and R_{11}, is equal to E_T. Using the law of proportionality,

$$E_{R_{10}} = E_T\frac{R_{10}}{R_{10} + R_{11}} = 50 \times \frac{2.97}{2.97 + 3.94} = 21.5 \text{ volts}$$

The voltage across R_5 is equal to the voltage across R_{10}. Thus,

$$I_{R_5} = E_{R_{10}}/R_5 = 21.5/5 = 4.3 \text{ amps}$$

1.13 Find the voltage across R_2 in Fig. 1.16.

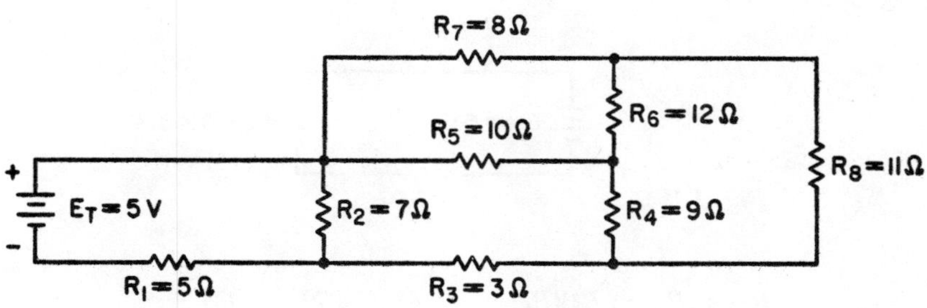

Fig. 1.16.

Answer

Convert the delta consisting of R_5, R_6, and R_7 to a wye (Fig. 1.16a).

$$R_9 = R_{ADJ}/R_T = R_6R_7/(R_5 + R_6 + R_7) = 12 \times 8/(10 + 12 + 8) = 3.2 \text{ ohms}$$

$$R_{10} = R_{ADJ}/R_T = R_5R_7/30 = (10 \times 8)/30 = 2.67 \text{ ohms}$$

$$R_{11} = R_{ADJ}/R_T = R_5R_6/30 = (10 \times 12)/30 = 4 \text{ ohms}$$

The series combination of R_9 and R_8 is in parallel with the series combination of R_{11} and R_4 to form R_{12} (Fig. 1.16b).

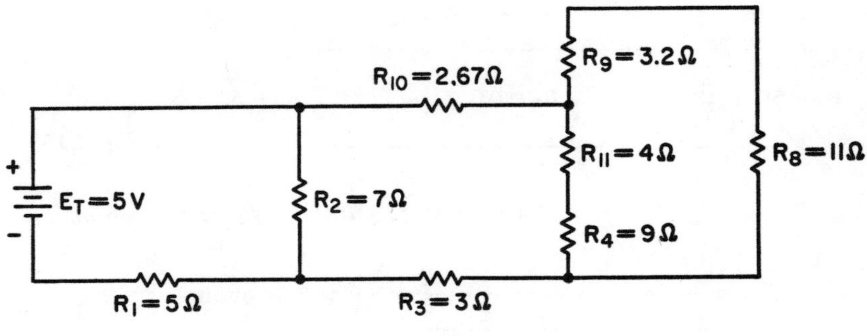

Fig. 1.16a.

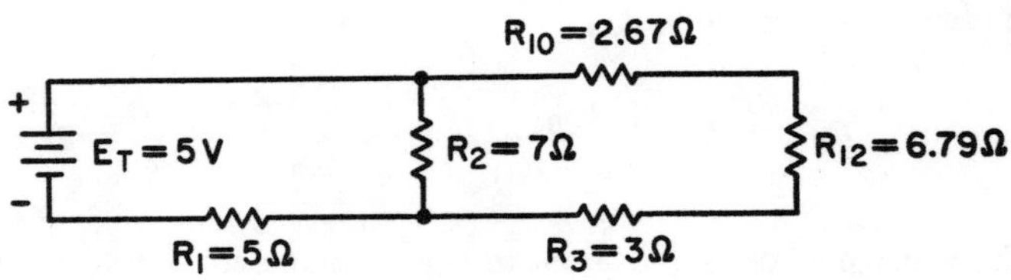

Fig. 1.16b.

$$R_{12} = \frac{(R_9 + R_8)(R_{11} + R_4)}{R_9 + R_8 + R_{11} + R_4} = \frac{(3.2 + 11)(4 + 9)}{3.2 + 11 + 4 + 9} = 6.79 \text{ ohms}$$

The series combination of R_{10}, R_{12}, and R_3 is in parallel with R_2 to form R_{13} (Fig. 1.16c).

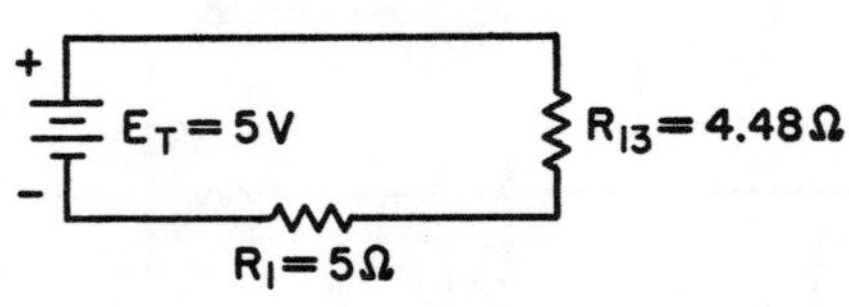

Fig. 1.16c.

$$R_{13} = \frac{(R_{10} + R_{12} + R_3)(R_2)}{R_{10} + R_{12} + R_3 + R_2} = \frac{(2.67 + 6.79 + 3)(7)}{2.67 + 6.79 + 3 + 7} = 4.48 \text{ ohms}$$

Using the law of proportionality,

$$E_{R13} = E_T \frac{R_{13}}{R_{13} + R_1} = 5 \times \frac{4.48}{4.48 + 5} = 2.36 \text{ volts}$$

Since the voltage across R_{13} is equal to that across R_2,

$$E_{R2} = 2.36 \text{ volts}$$

2

Thevenin's Theorem

A second method used to reduce a complex circuit to a simple circuit is called Thevenin's theorem. Thevenin's theorem states that any two-terminal network can be reduced to a voltage source, $E_{Thevenin}$, and a series resistance, $R_{Thevenin}$. $R_{Thevenin}$ will be written R_{Thev}; $E_{Thevenin}$ will be written E_{Thev}.

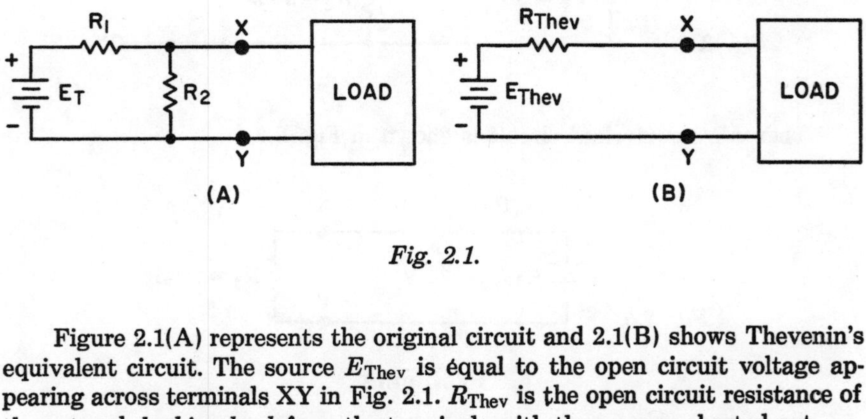

Fig. 2.1.

Figure 2.1(A) represents the original circuit and 2.1(B) shows Thevenin's equivalent circuit. The source E_{Thev} is équal to the open circuit voltage appearing across terminals XY in Fig. 2.1. R_{Thev} is the open circuit resistance of the network *looking back* from the terminals with the sources shorted out.

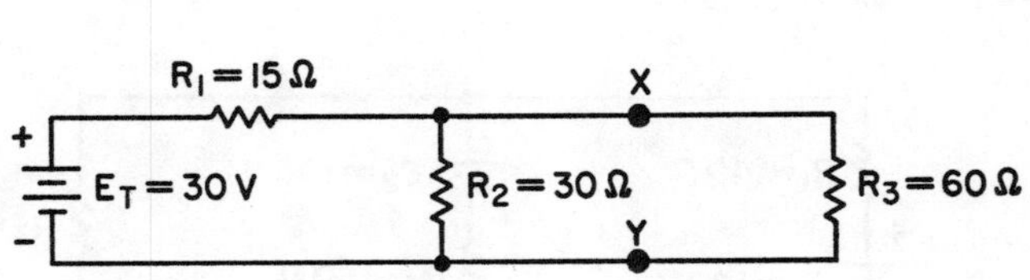

Fig. 2.2.

Example

Find Thevenin's equivalent circuit at terminals XY in Fig. 2.2.

First find R_{Thev} by removing the load and shorting out the source (see Fig. 2.3):

$$R_{Thev} = R_1R_2/(R_1 + R_2) = (15 \times 30)/(15 + 30) = 10 \text{ ohms}$$

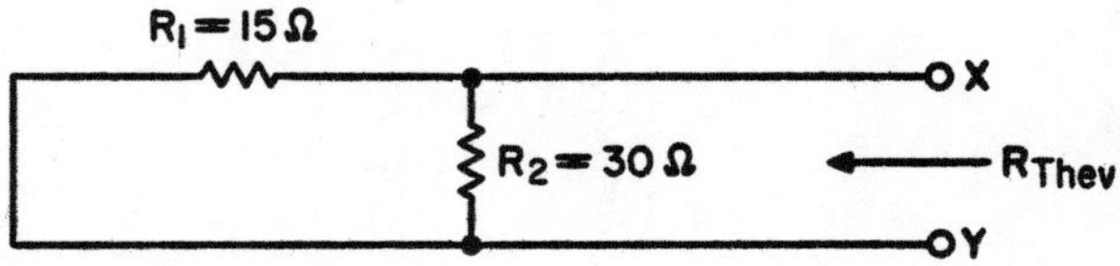

Fig. 2.3.

Now find E_{Thev} by removing the load and finding the open circuit voltage (see Fig. 2.4). Using the law of proportionality,

$$E_{R2} = E_{Thev} = E_T \frac{R_2}{R_1 + R_2} = 30 \times \frac{30}{45} = 20 \text{ volts}$$

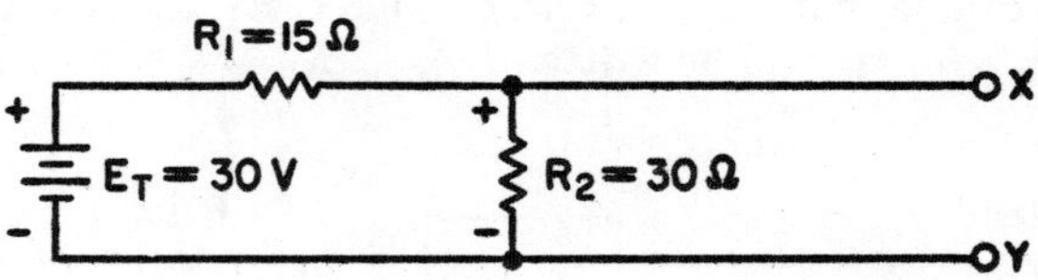

Fig. 2.4.

Thevenin's equivalent circuit is shown in Fig. 2.5.

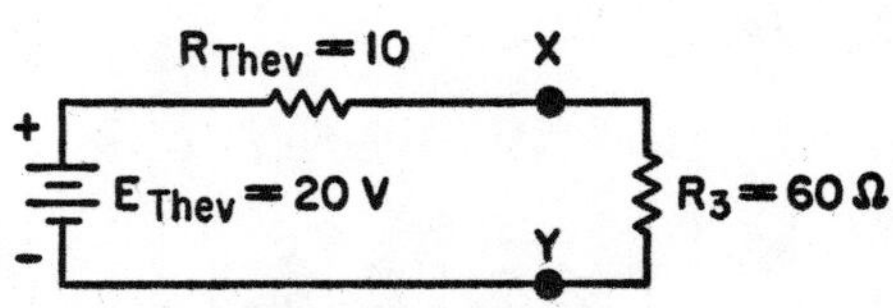

Fig. 2.5.

Problems

2.1 Find the current flowing through R_3 in Fig. 2.6.

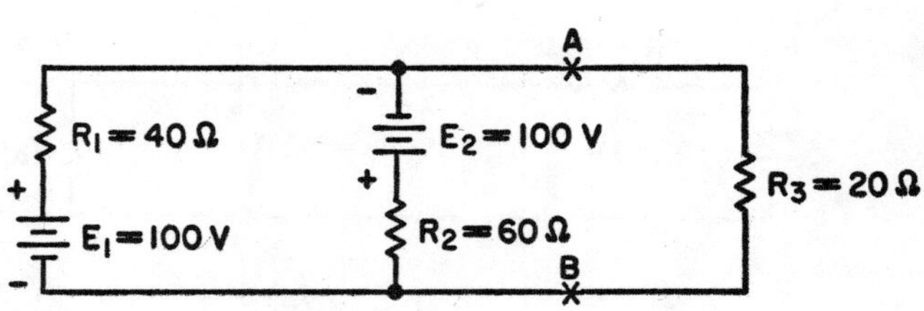

Fig. 2.6.

Answer

Open the circuit at points A and B and Thevenize (Fig. 2.6a).

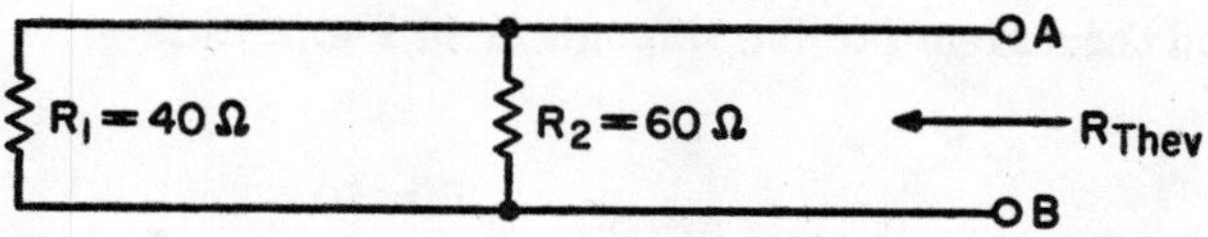

Fig. 2.6a.

$$R_{\text{Thev}} = R_1R_2/(R_1 + R_2) = (40 \times 60)/(40 + 60) = 24 \text{ ohms}$$

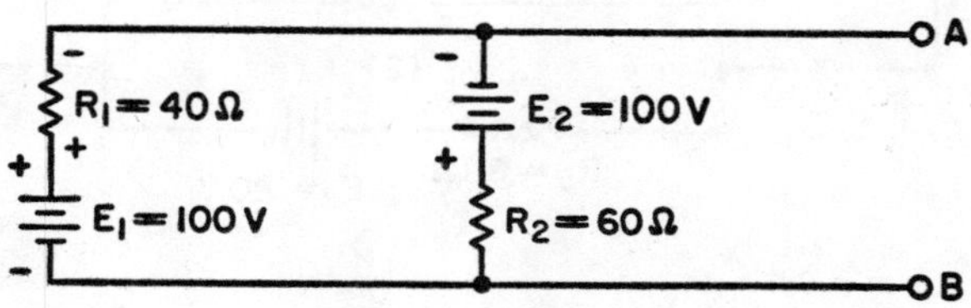

Fig. 2.6b.

To find E_{Thev} we must first find E_{R_1} (or E_{R_2}). E_1 and E_2 are series aiding; therefore,

$$E_{\text{Loop}} = E_1 + E_2 = 100 + 100 = 200 \text{ volts}$$

Using the law of proportionality,

$$E_{R_1} = E_{\text{Loop}} \frac{R_1}{R_1 + R_2} = 200 \times \frac{40}{100} = 80 \text{ volts}$$

The voltage across R_1 is opposing E_1; therefore,

$$E_{\text{Thev}} = E_1 - E_{R_1} = 100 - 80 = 20 \text{ volts}$$

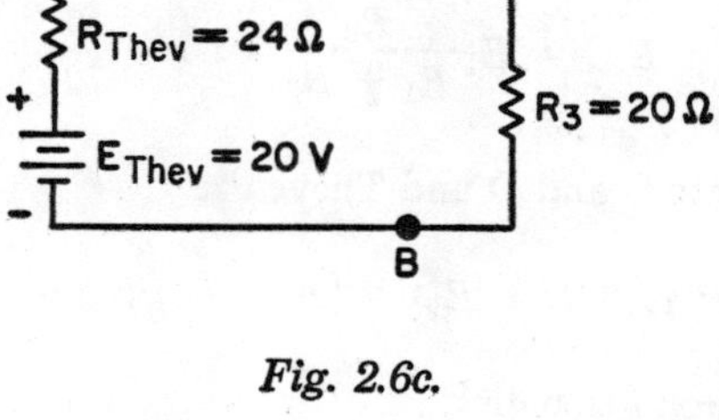

Fig. 2.6c.

$$I_{R_3} = E_{\text{Thev}}/(R_{\text{Thev}} + R_3) = 20/(24 + 20) = 20/44$$
$$= 0.455 \text{ amp flowing from B to A}$$

2.2 Find the current flowing through R_6 in Fig. 2.7.

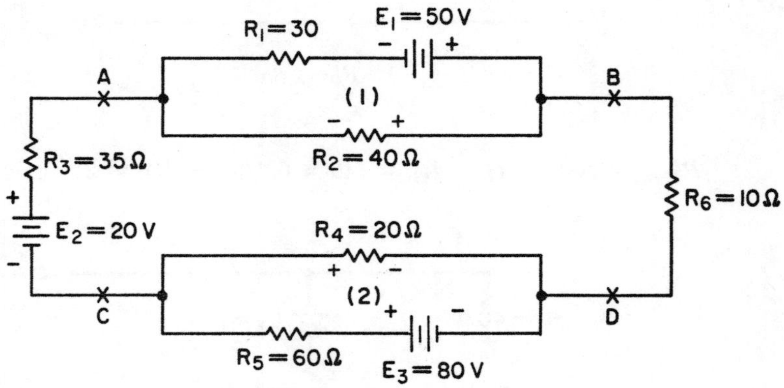

Fig. 2.7.

Answer

To simplify the above circuit, Thevenize loop 1 and loop 2 (Fig. 2.7a).

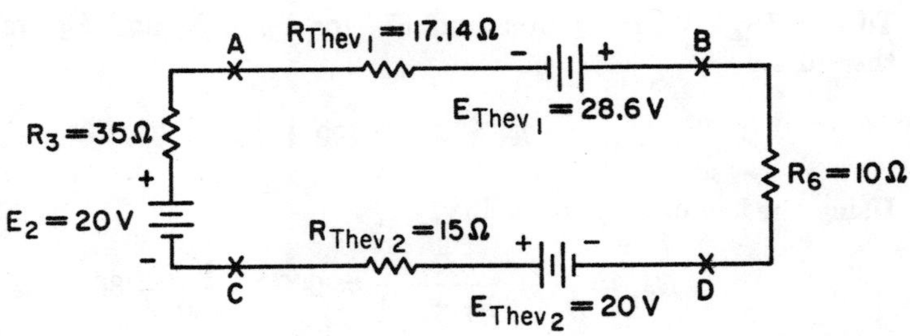

Fig. 2.7a.

Open loop 1 at points A and B and Thevenize:

$$R_{\text{Thev 1}} = R_1R_2/(R_1 + R_2) = (30 \times 40)/(30 + 40) = 17.14 \text{ ohms}$$

Using the law of proportionality,

$$E_{\text{Thev 1}} = E_{R_2} = E_1\frac{R_2}{R_1 + R_2} = 50 \times \frac{40}{70} = 28.6 \text{ volts}$$

Open loop 2 at points C and D and Thevenize:

$$R_{\text{Thev 2}} = R_4R_5/(R_4 + R_5) = (20 \times 60)/(20 + 60) = 15 \text{ ohms}$$

Using the law of proportionality,

$$E_{\text{Thev 2}} = E_{R_4} = E_3\frac{R_4}{R_4 + R_5} = 80 \times \frac{20}{80} = 20 \text{ volts}$$

Figure 2.7a shows a series circuit; the current through R_6 can therefore be determined by using Ohm's law. Note that all three batteries are connected series aiding.

$$I_{R_6} = \frac{E_{\text{Thev 1}} + E_2 + E_{\text{Thev 2}}}{R_{\text{Thev 1}} + R_3 + R_{\text{Thev 2}} + R_6} = \frac{(28.6 + 20 + 20)}{(17.14 + 35 + 15 + 10)}$$

$I_{R_6} = 0.89$ amp flowing from D to B

2.3 What is the equivalent E_{Thev} and R_{Thev} across points X and Y in Fig. 2.8?

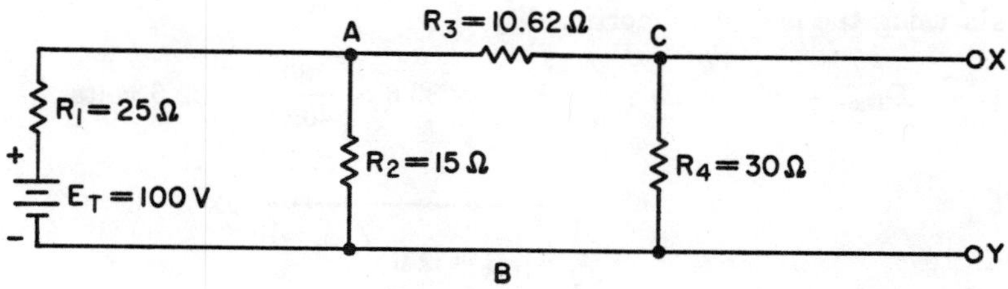

Fig. 2.8.

Answer

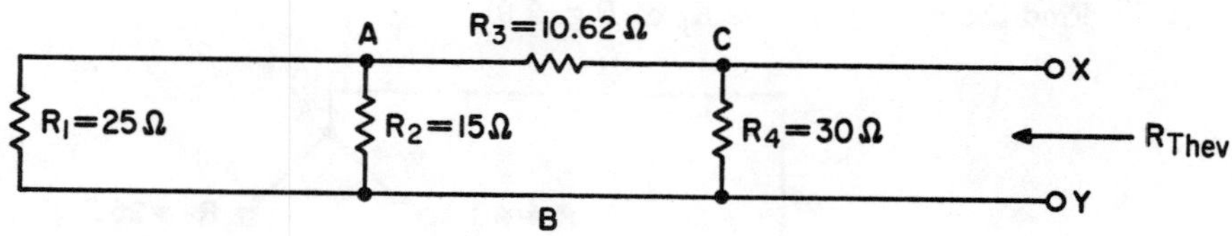

Fig. 2.8a.

Resistors R_1 and R_2 are in parallel with each other (Fig. 2.8a) to form R_{AB}:

$$R_{AB} = R_1R_2/(R_1 + R_2) = (25 \times 15)/(25 + 15) = 9.38 \text{ ohms}$$

The series combination of R_{AB} and R_3 is in parallel with R_4 to form R_{Thev}.

$$R_{\text{Thev}} = \frac{(R_{AB} + R_3)R_4}{R_{AB} + R_3 + R_4} = \frac{(9.38 + 10.62)30}{9.38 + 10.62 + 30} = 12 \text{ ohms}$$

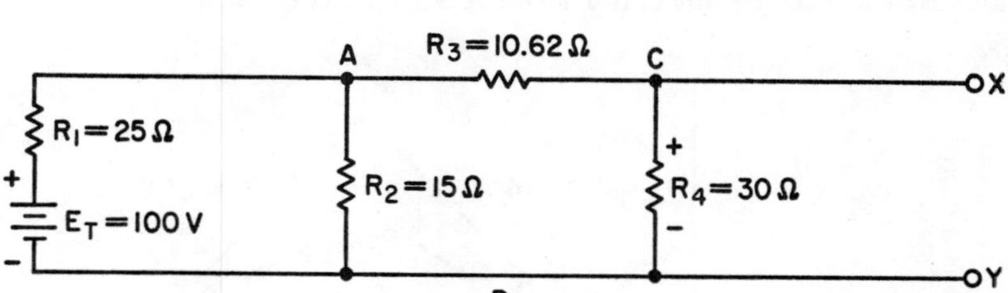

Fig. 2.8b.

E_{Thev} is equal to the voltage dropped across R_4 (Fig. 2.8b). The series combination of R_3 and R_4 is in parallel with R_2 to form R_{AB}:

$$R_{AB} = \frac{(R_3 + R_4)R_2}{R_3 + R_4 + R_2} = \frac{(30 + 10.62)15}{30 + 10.62 + 15} = 11 \text{ ohms}$$

You now have a series circuit consisting of E_T, R_1, and R_{AB}. Using the law of proportionality,

$$E_{AB} = E_T \frac{R_{AB}}{R_1 + R_{AB}} = 100 \times \frac{11}{36} = 30.6 \text{ volts} = E_{R_3} + E_{R_4}$$

Again using the law of proportionality,

$$E_{Thev} = E_{R_4} = E_{AB} \frac{R_4}{R_3 + R_4} = 30.6 \times \frac{30}{40.62} = 22.6 \text{ volts}$$

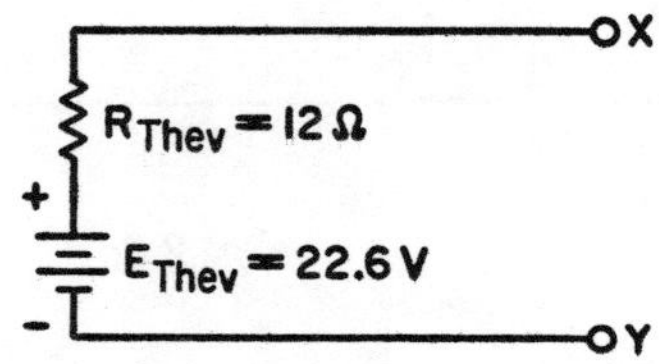

Fig. 2.8c.

2.4 Find the voltage across R_5 of Fig. 2.9.

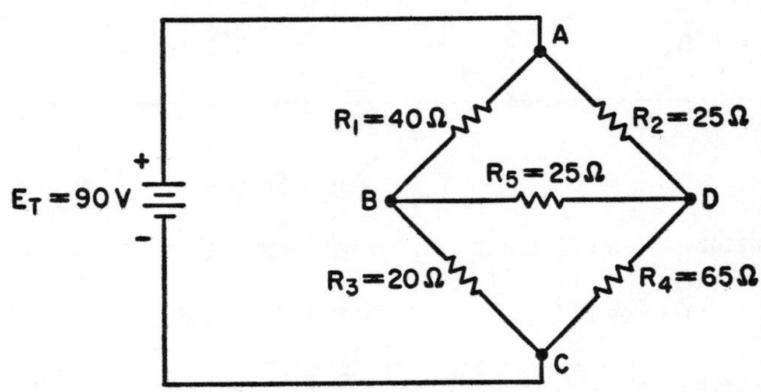

Fig. 2.9.

Answer

Remove R_5 from the circuit and Thevenize Fig. 2.9 (Fig. 2.9a).

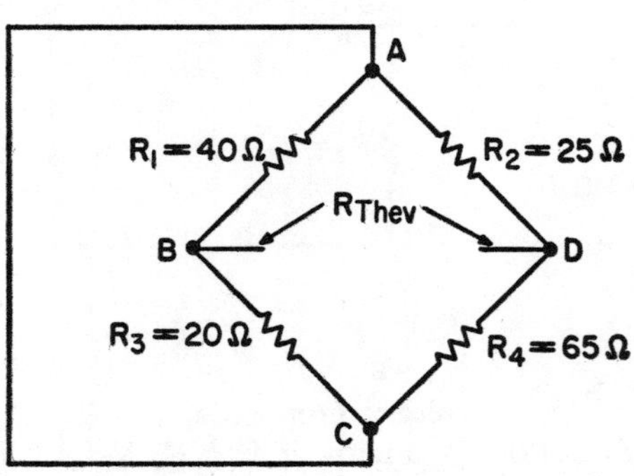

Fig. 2.9a.

R_{Thev} is the resistance between points B and D. Since the battery is shorted out, points A and C are the same and R_{Thev} is equal to the parallel combination of R_1 and R_3 in series with the parallel combination of R_2 and R_4.

$$R_{Thev} = \frac{R_1R_3}{R_1 + R_3} + \frac{R_2R_4}{R_2 + R_4} = \frac{40 \times 20}{40 + 20} + \frac{25 \times 65}{25 + 65}$$

$$R_{Thev} = 13.3 + 18.1 = 31.4 \text{ ohms}$$

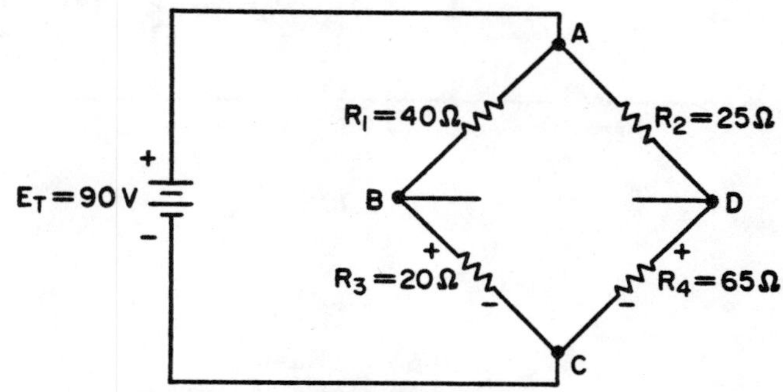

Fig. 2.9b.

E_{Thev} is the open circuit voltage across points B and D. Using the law of proportionality,

$$E_{R_3} = E_{BC} = E_T\frac{R_3}{R_1 + R_3} = 90 \times \frac{20}{60} = 30 \text{ volts}$$

$$E_{R_4} = E_{DC} = E_T\frac{R_4}{R_2 + R_4} = 90 \times \frac{65}{90} = 65 \text{ volts}$$

$$E_{Thev} = E_{DB} = E_{DC} - E_{BC} = 65 - 30 = 35 \text{ volts}$$

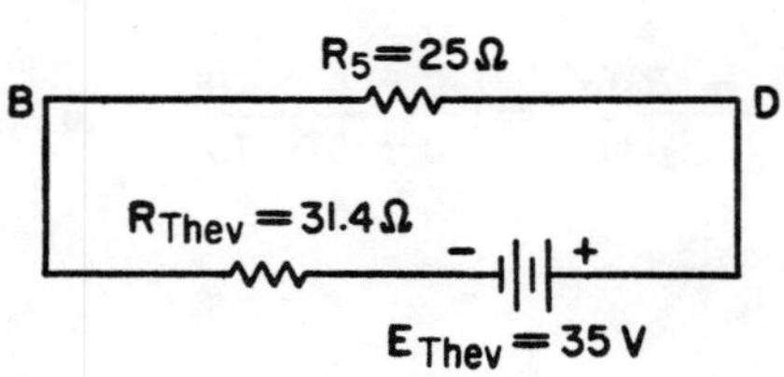

Fig. 2.9c.

Using the law of proportionality,

$$E_{R_5} = E_{Thev}\frac{R_5}{R_5 + R_{Thev}} = 35 \times \frac{25}{56.4} = 15.5 \text{ volts}$$

2.5 Find the current flowing through R_5 in Fig. 2.10.

Answer

To simplify the above circuit, Thevenize loop 1 and loop 2 (Fig. 2.10a).

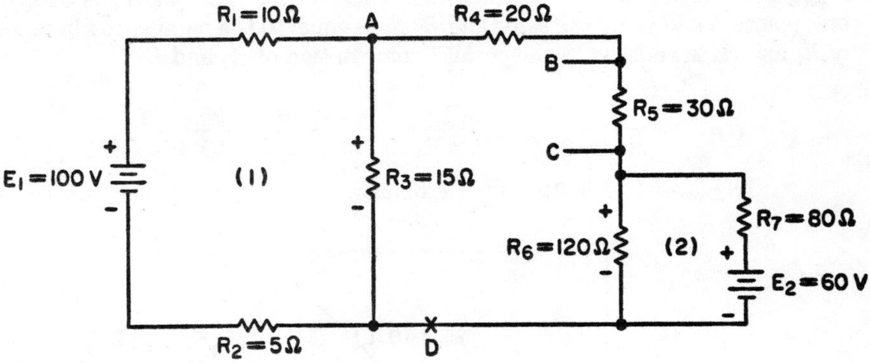

Fig. 2.10.

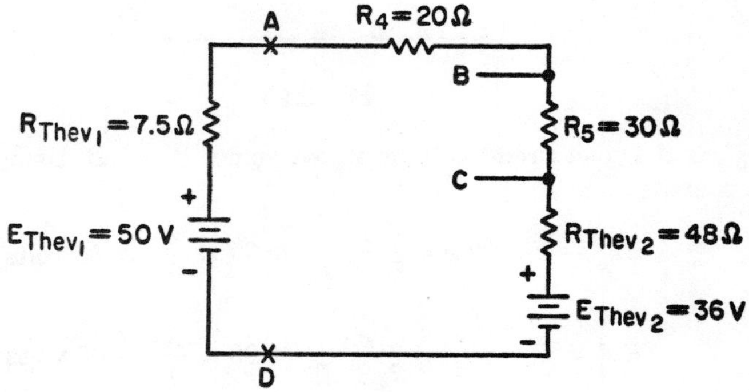

Fig. 2.10a

Open the circuit at points A and D and Thevenize loop 1.

$$R_{\text{Thev }1} = \frac{(R_1 + R_2)R_3}{R_1 + R_2 + R_3} = \frac{(10 + 5) \times 15}{10 + 5 + 15} = 7.5 \text{ ohms}$$

Using the law of proportionality

$$E_{\text{Thev }1} = E_{R_3} = E_1 \frac{R_3}{R_1 + R_2 + R_3} = 100 \times \frac{15}{30} = 50 \text{ volts}$$

Open the circuit at points C and D and Thevenize loop 2.

$$R_{\text{Thev }2} = R_6R_7/(R_6 + R_7) = (120 \times 80)/(120 + 80) = 48 \text{ ohms}$$

Using the law of proportionality,

$$E_{\text{Thev }2} = E_{R_6} = E_2 \frac{R_6}{R_6 + R_7} = 60 \times \frac{120}{200} = 36 \text{ volts}$$

Figure 2.10a shows a series circuit; therefore, the current through R_5 can now be determined by using Ohm's law. Note that $E_{\text{Thev 1}}$ and $E_{\text{Thev 2}}$ are connected series opposing.

$$I_{R_5} = \frac{E_{\text{Thev 1}} - E_{\text{Thev 2}}}{R_{\text{Thev 1}} + R_4 + \text{R}_5 + R_{\text{Thev 2}}} = \frac{(50 - 36)}{(7.5 + 20 + 30 + 48)}$$

$$I_{R_5} = 0.133 \text{ amp flowing from C to B}$$

2.6 Using Thevenim's theorem, find e_o of Fig. 2.11.

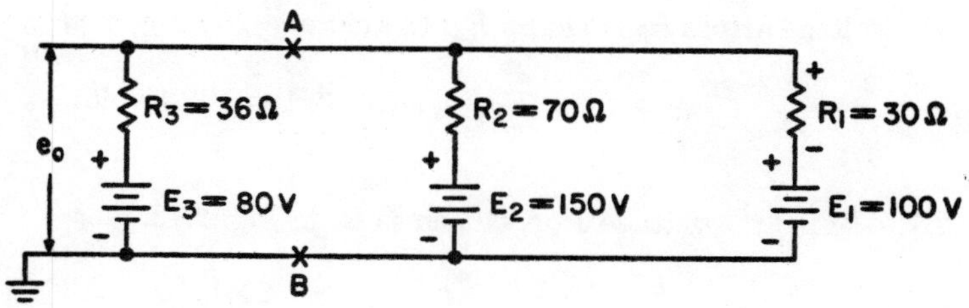

Fig. 2.11.

Answer

Open the circuit at points A and B, and Thevenize the circuit to the right of these two points (Fig. 2.11a).

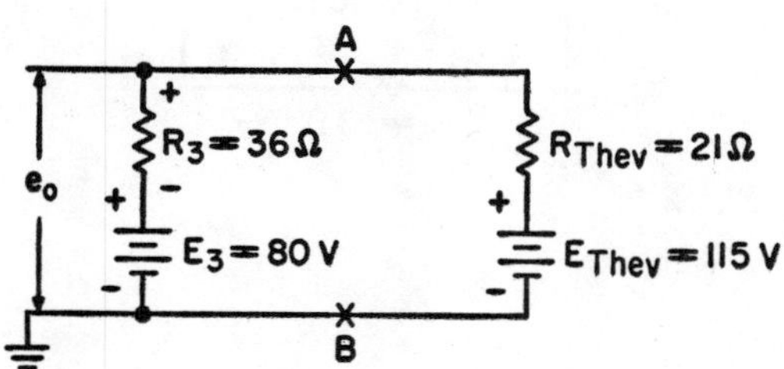

Fig. 2.11a.

$$R_{\text{Thev}} = R_1R_2/(R_1 + R_2) = (30 \times 70)/(30 + 70) = 21 \text{ ohms}$$

To find E_{Thev} we must first find E_{R_1} (or E_{R_2}). E_1 and E_2 are connected series opposing; therefore,

$$E_{\text{Loop}} = E_2 - E_1 = 150 - 100 = 50 \text{ volts}$$

Using the law of proportionality,

$$E_{R_1} = E_{\text{Loop}} \frac{R_1}{R_1 + R_2} = 50 \times \frac{30}{100} = 15 \text{ volts}$$

The voltage across R_1 is aiding E_1; therefore,

$$E_{\text{Thev}} = E_1 + E_{R_1} = 100 + 15 = 115 \text{ volts}$$

The voltage e_o is equal to the algebraic sum of the voltage across R_3 and E_3 (or the voltage across R_{Thev} and E_{Thev}). In Fig. 2.11a, E_3 and E_{Thev} are connected series opposing; therefore,

$$E_{\text{Loop}} = E_{\text{Thev}} - E_3 = 115 - 80 = 35 \text{ volts}$$

Using the law of proportionality,

$$E_{R3} = E_{\text{Loop}} \frac{R_3}{R_3 + R_{\text{Thev}}} = 35 \times \frac{36}{57} = 22.1 \text{ volts}$$

The voltage across R_3 is aiding E_3; therefore,

$$e_o = E_3 + E_{R3} = 80 + 22.1 = 102.1 \text{ volts}$$

2.7 Find the voltage dropped across R_4 in Fig. 2.12.

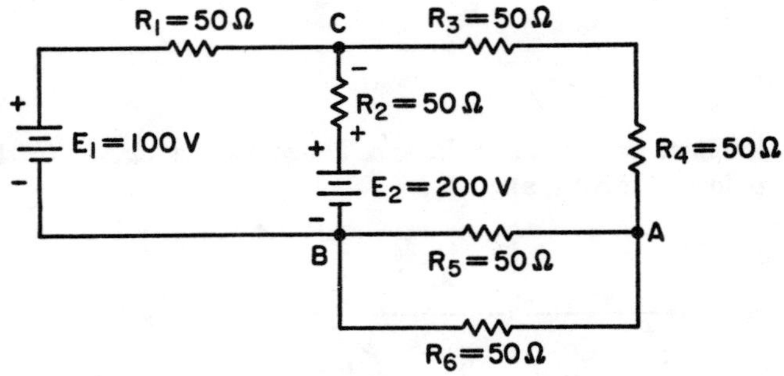

Fig. 2.12.

Answer

To find the voltage across R_4, open the circuit at points C and B and Thevenize the circuit to the left of these points (Fig. 2.12a). Resistors R_5 and R_6, which are in parallel with each other, can be replaced by resistor R_{AB}.

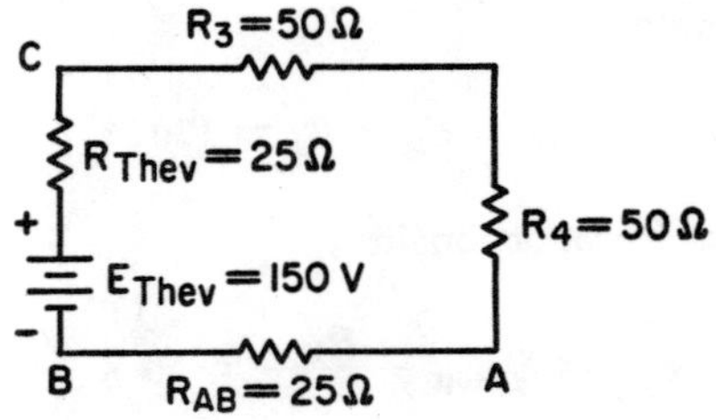

Fig. 2.12a.

$$R_{AB} = \frac{R_5 R_6}{R_5 + R_6} = \frac{50 \times 50}{50 + 50} = 25 \text{ ohms}$$

$$R_{\text{Thev}} = \frac{R_1R_2}{R_1 + R_2} = \frac{50 \times 50}{50 + 50} = 25 \text{ ohms}$$

To find E_{Thev} we must first find E_{R2} (or E_{R1}). E_1 and E_2 are connected series opposing; therefore,

$$E_{\text{Loop}} = E_2 - E_1 = 200 - 100 = 100 \text{ volts}$$

Using the law of proportionality,

$$E_{R2} = E_{\text{Loop}} \frac{R_2}{R_1 + R_2} = 100 \times \frac{50}{100} = 50 \text{ volts}$$

The voltage across R_2 is opposing E_2; therefore,

$$E_{\text{Thev}} = E_2 - E_{R2} = 200 - 50 = 150 \text{ volts}$$

Figure 2.12a shows a series circuit. Using the law of proportionality,

$$E_{R4} = E_{\text{Thev}} \frac{R_4}{R_{\text{Thev}} + R_3 + R_4 + R_{\text{AB}}} = 150 \times \frac{50}{150} = 50 \text{ volts}$$

2.8 Find the voltage dropped across R_7 in Fig. 2.13.

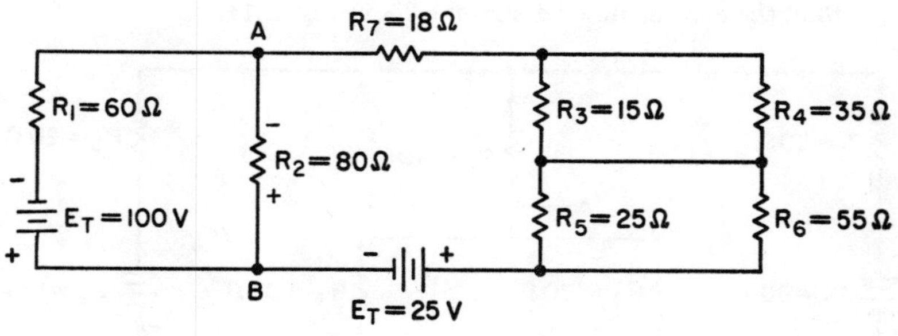

Fig. 2.13.

Answer

To find the voltage across R_7, open the circuit at points A and B and Thevenize the circuit to the left of these points (Fig. 2.13a). The parallel combination of R_3 and R_4 is in series with the parallel combination of R_5 and R_6 to form R_8.

$$R_8 = \frac{R_3R_4}{R_3 + R_4} + \frac{R_5R_6}{R_5 + R_6} = \frac{15 \times 35}{15 + 35} + \frac{25 \times 55}{25 + 55} = 10.5 + 17.2$$
$$= 27.7 \text{ ohms}$$

$$R_{\text{Thev}} = R_1R_2/(R_1 + R_2) = (60 \times 80)/(60 + 80) = 34.3 \text{ ohms}$$

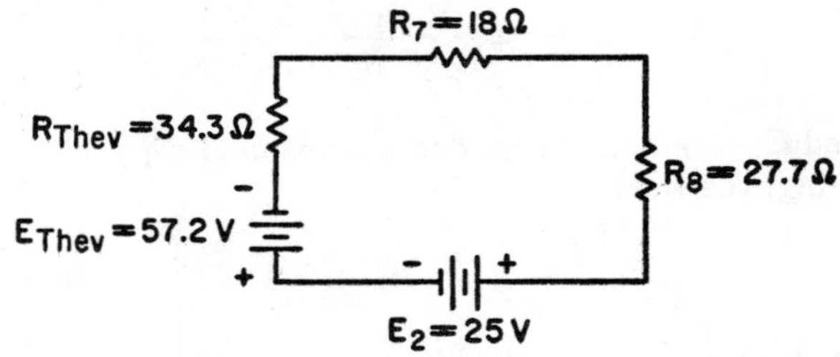

Fig. 2.13a.

To find E_{Thev}, we must first find the voltage across R_2. Using the law of proportionality,

$$E_{Thev} = E_{R_2} = E_1 \frac{R_2}{R_1 + R_2} = 100 \times \frac{80}{140} = 57.1 \text{ volts}$$

Figure 2.13a shows a series circuit. Note that E_{Thev} and E_2 are connected series aiding. Using the law of proportionality,

$$E_{R_7} = (E_{Thev} + E_2) \frac{R_7}{R_{Thev} + R_7 + R_8} = 82.1 \times \frac{18}{80} = 18.5 \text{ volts}$$

2.9 Find the current flowing through R_2 in Fig. 2.14.

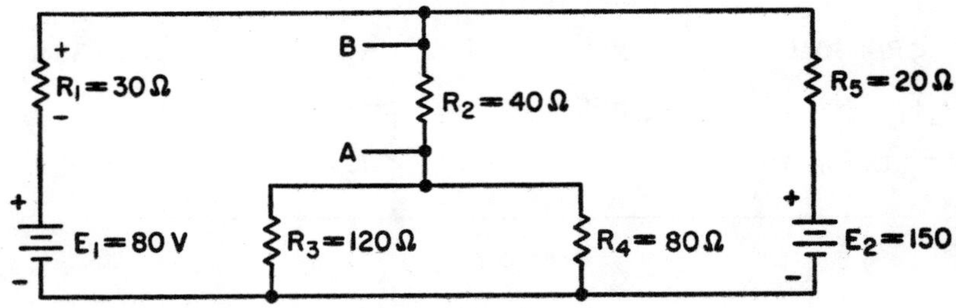

Fig. 2.14.

Answer

Replace the two branches consisting of R_1 and E_1 and R_5 and E_2 with their Thevenin equivalents (Fig. 2.14a). The parallel combination of R_3 and R_4 forms R_6.

$$R_6 = R_3R_4/(R_3 + R_4) = (120 \times 80)/(120 + 80) = 48 \text{ ohms}$$

$$R_{Thev} = R_1R_5/(R_1 + R_5) = (30 \times 20)/(30 + 20) = 12 \text{ ohms}$$

To find E_{Thev} we must first find E_{R_1} (or E_{R_5}). E_1 and E_2 are connected series opposing.

$$E_{Loop} = E_2 - E_1 = 150 - 80 = 70 \text{ volts}$$

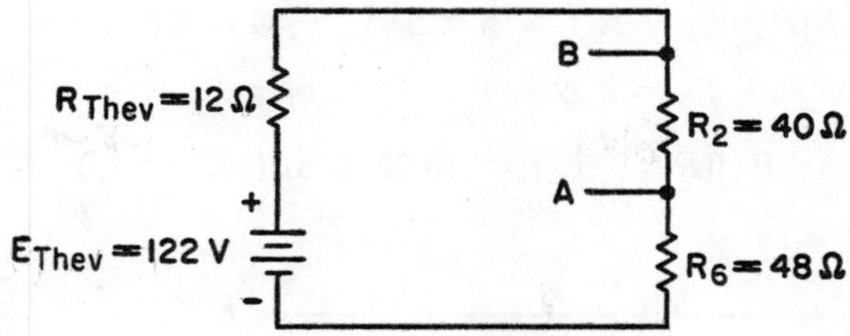

Fig. 2.14a.

Using the law of proportionality,

$$E_{R_1} = E_{\text{Loop}}\frac{R_1}{R_1 + R_5} = 70 \times \frac{30}{50} = 42 \text{ volts}$$

The voltage across R_1 is aiding E_1.

$$E_{\text{Thev}} = E_1 + E_{R_1} = 80 + 42 = 122 \text{ volts}$$

Figure 2.14a shows a series circuit; therefore, the current flowing through R_2 can now be determined by Ohm's law.

$$I_{R_2} = \frac{E_{\text{Thev}}}{R_{\text{Thev}} + R_2 + R_6} = \frac{122}{(12 + 40 + 48)}$$

$$I_{R_2} = 1.22 \text{ amps flowing from A to B}$$

2.10 Find the voltage across R_4 in Fig. 2.15.

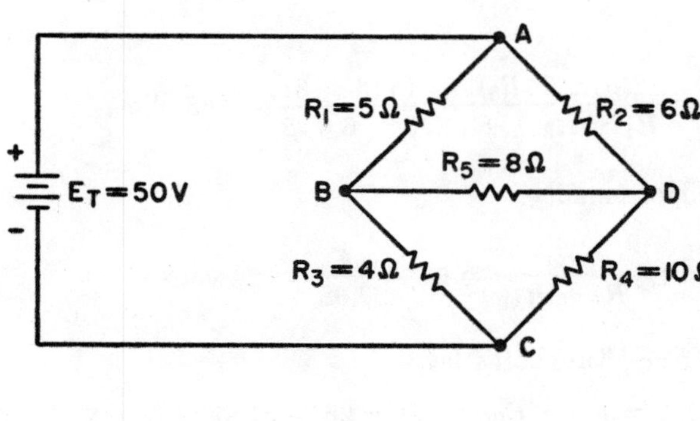

Fig. 2.15.

Answer

R_{Thev} is the resistance between points C and D (Fig. 2.15a). Since the battery is shorted out, points A and C are the same. R_{Thev} is therefore equal to the parallel combination of R_1 *and* R_3 (forming R_{T_1}), which is in series with R_5 (forming R_{T_2}) and this combination is in parallel with R_2 to form R_{Thev}.

$$R_{T_1} = R_1R_3/(R_1 + R_3) = (5 \times 4)/(5 + 4) = 2.22 \text{ ohms}$$

$$R_{T_2} = R_{T_1} + R_5 = 2.22 + 8 = 10.22 \text{ ohms}$$

$$R_{\text{Thev}} = R_{T2}R_2/(R_{T2} + R_2) = (10.22 \times 6)/(10.22 + 6) = 3.78 \text{ ohms}$$

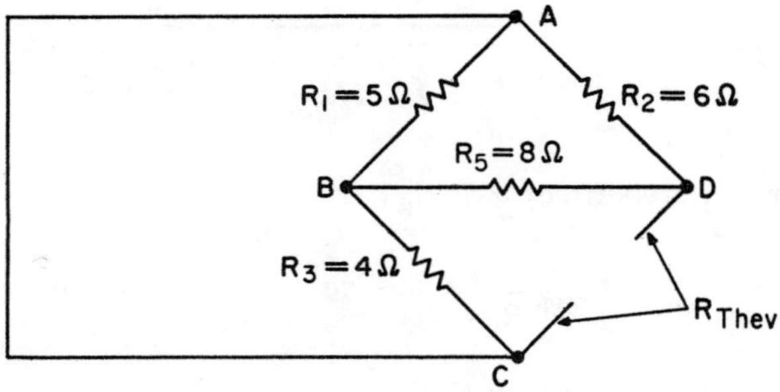

Fig. 2.15a.

E_{Thev} is the open circuit voltage across points D and C (Fig. 2.15b). The series combination of R_2 and R_5 is in parallel with R_1 to form R_{AB}.

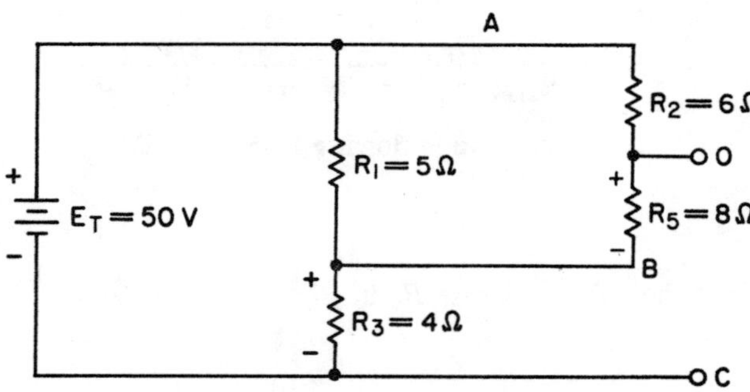

Fig. 2.15b.

$$R_{AB} = \frac{R_1(R_2 + R_5)}{R_1 + R_2 + R_5} = \frac{(5)(6 + 8)}{5 + 6 + 8} = 3.68 \text{ ohms}$$

Using the law of proportionality,

$$E_{BC} = E_T \frac{R_3}{R_3 + R_{AB}} = 50 \times \frac{4}{7.68} = 26 \text{ volts}$$

Since $E_T = E_{AB} + E_{BC}$ (Kirchhoff's law),

$$E_{AB} = E_T - E_{BC} = 50 - 26 = 24 \text{ volts}$$

Again using the law of proportionality,

$$E_{DB} = E_{AB} \frac{R_5}{R_2 + R_5} = 24 \times \frac{8}{14} = 13.7 \text{ volts}$$

$$E_{Thev} = E_{DC} = E_{DB} + E_{BC} = 13.7 + 26 = 39.7 \text{ volts}$$

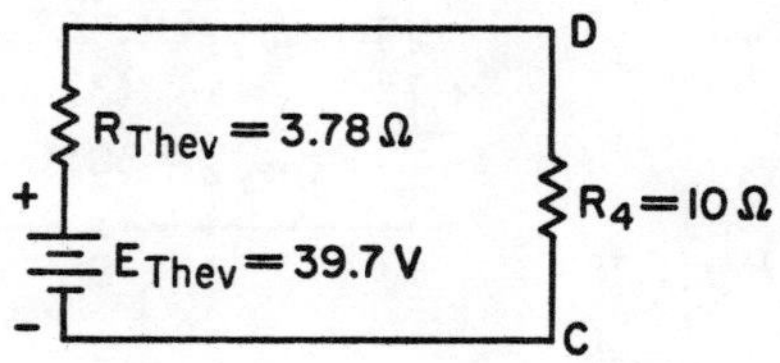

Fig. 2.15c.

Using the law of proportionality,

$$E_{R4} = E_{Thev}\frac{R_4}{R_{Thev} + R_4} = 39.7 \times \frac{10}{13.78} = 28.8 \text{ volts}$$

2.11 Find the current flowing through R_4 in Fig. 2.16.

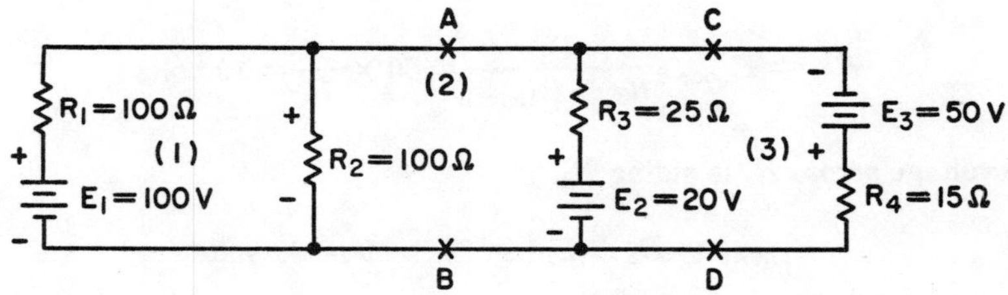

Fig. 2.16.

Answer

Open the circuit at points A and B and Thevenize loop 1 (Fig. 2.16a).

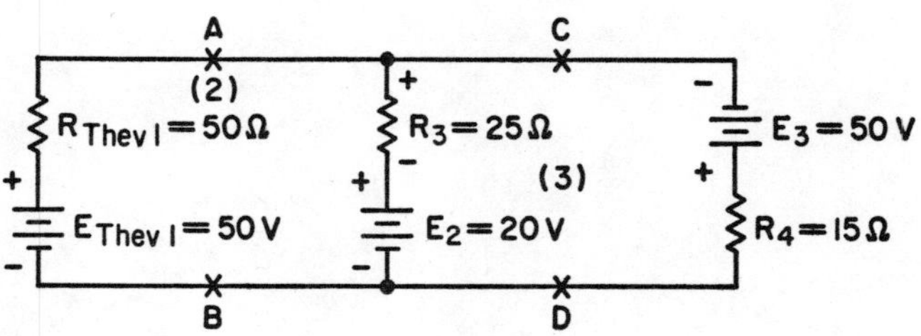

Fig. 2.16a.

$$R_{Thev\ 1} = R_1R_2/(R_1 + R_2) = (100 \times 100)/(100 + 100) = 50 \text{ ohms}$$

To find $E_{Thev\ 1}$ we must find the voltage across R_2.

$$E_{Thev\ 1} = E_{R2} = E_1\frac{R_1}{R_1 + R_2} = 100 \times \frac{100}{200} = 50 \text{ volts}$$

Using Fig. 2.16a, open the circuit at points C and D and Thevenize loop 2 (Fig. 2.16b).

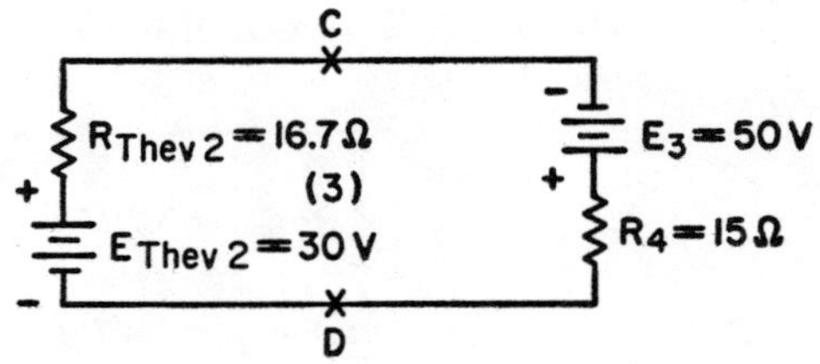

Fig. 2.16b.

$R_{\text{Thev 2}} = (R_{\text{Thev 1}} \times R_3)/(R_{\text{Thev 1}} + R_3) = (50 \times 25)/(50 + 25) = 16.7$ ohms

To find E_{Thev_2}, we must first find E_{R_3} (or $E_{R_{\text{Thev 1}}}$). $E_{\text{Thev 1}}$ and E_2 are connected series opposing.

$$E_{\text{Loop 2}} = E_{\text{Thev 1}} - E_2 = 50 - 20 = 30 \text{ volts}$$

Using the law of proportionality,

$$E_{R_3} = E_{\text{Loop 2}} \frac{R_3}{R_3 + R_{\text{Thev 1}}} = 30 \times \frac{25}{75} = 10 \text{ volts}$$

The voltage across R_3 is aiding E_2.

$$E_{\text{Thev 2}} = E_2 + E_{R_3} = 20 + 10 = 30 \text{ volts}$$

Figure 2.16b shows a series circuit; the current flowing through R_4 can therefore be determined by Ohm's law. Note that E_{Thev2} and E_3 are connected series aiding.

$$I_{R_4} = (E_{\text{Thev 2}} + E_3)/(R_{\text{Thev 2}} + R_4) = 80/31.7$$
$$= 2.52 \text{ amps flowing from point D to C}$$

3

Norton's Theorem

Another method used to reduce a complex circuit to a simple circuit is called Norton's theorem. Norton's theorem states that any two-terminal network can be reduced to a constant current source, I_{Norton}, and a parallel equivalent resistance, R_{Norton}. I_{Norton} will be written I_N; R_{Norton} will be written R_N. A circle containing an arrow denotes the constant current source, with the arrow indicating the direction of current (electron flow).

In Fig. 3.1, (A) represents the original circuit and (B) shows Norton's equivalent circuit. The source I_N is equal to the short circuit current appearing across the terminals XY. R_N is the open circuit resistance of the network looking back from the terminals with the sources shorted out. You will notice that R_N is the same as R_{Thev}. The only difference is that R_N is in parallel with the source.

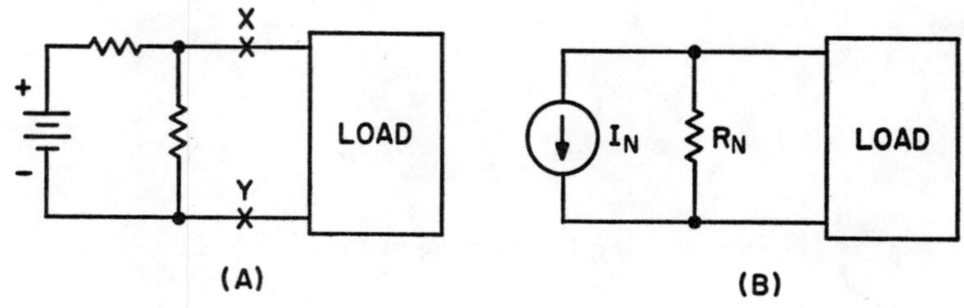

Fig. 3.1.

Example

In Fig. 3.2, find Norton's equivalent circuit.

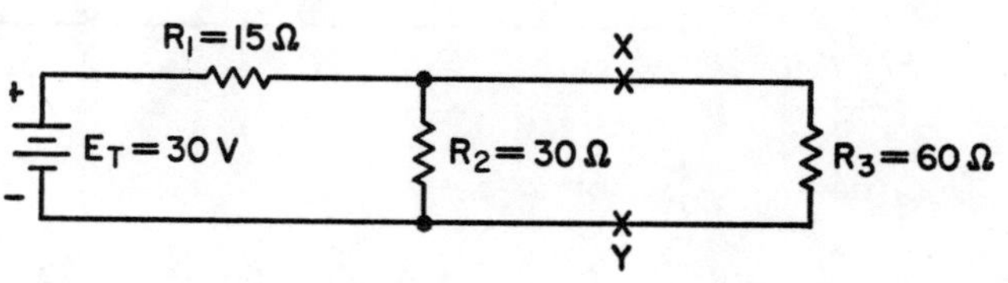

Fig. 3.2.

First find the short circuit current (Fig. 3.3).

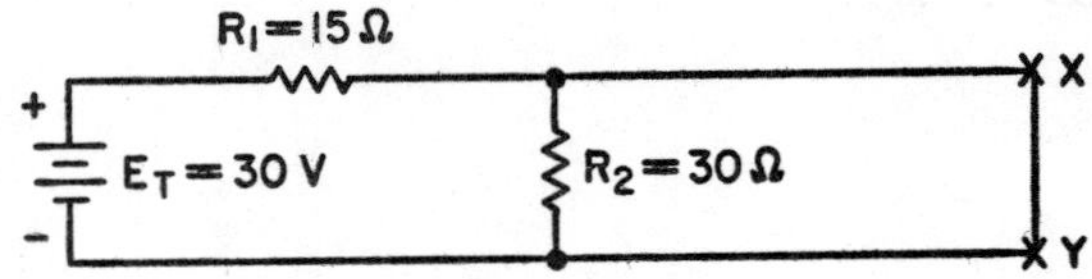

Fig. 3.3.

$$I_N = E_T/R_1 = 30/15 = 2 \text{ amps}$$

Then, with the source shorted out (Fig. 3.4), find R_N.

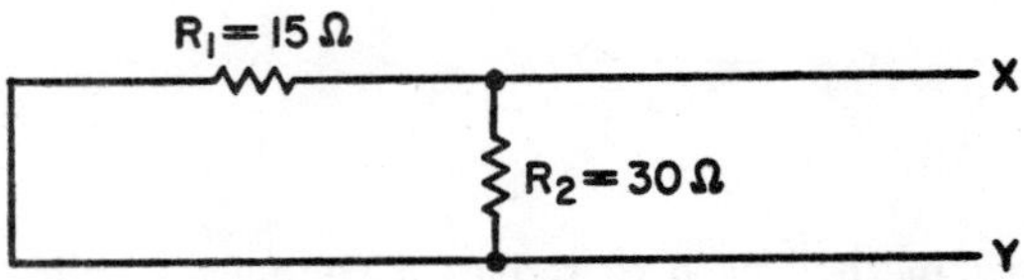

Fig. 3.4.

$$R_N = R_1R_2/(R_1 + R_2) = (15 \times 30)/(15 + 30) = 10 \text{ ohms}$$

Norton's equivalent circuit for Fig. 3.2 is given by Fig. 3.5.

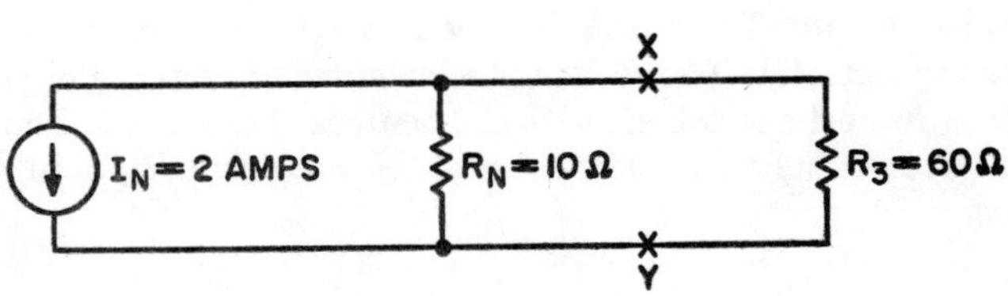

Fig. 3.5.

Problems

3.1 Find the current flowing through R_3 in Fig. 3.6.

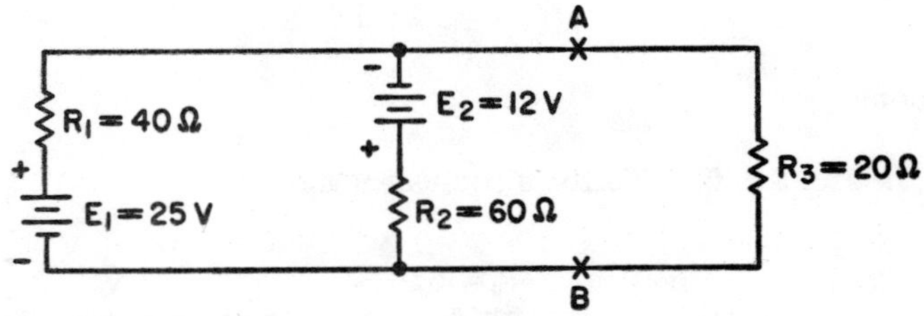

Fig. 3.6.

Answer

$$R_N = R_1R_2/(R_1 + R_2) = (40 \times 60)/(40 + 60) = 24 \text{ ohms}$$

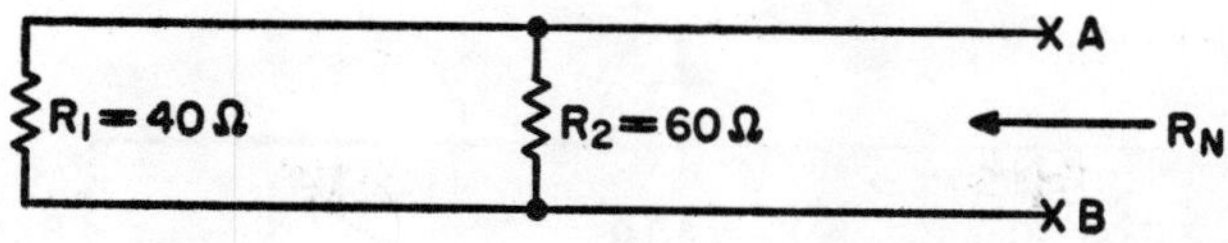

Fig. 3.6a.

$I_{N1} = E_1/R_1 = 25/40 = 0.625$ **amp flowing from B to A**

$I_{N2} = E_2/R_2 = 12/60 = 0.2$ **amp flowing from A to B**

$I_{NT} = I_{N1} - I_{N2} = 0.625 - 0.2 = 0.425$ **amp flowing from B to A**

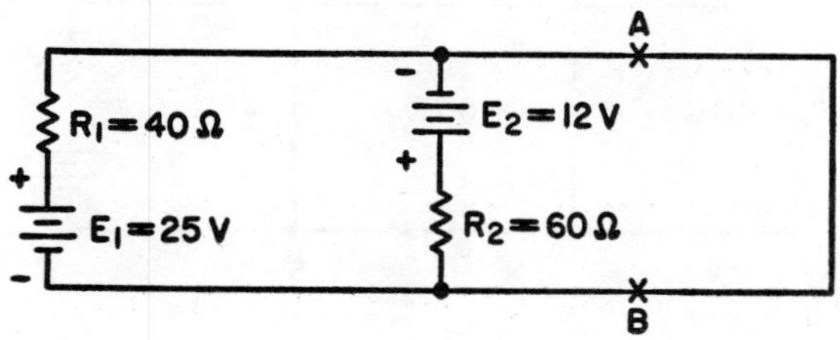

Fig. 3.6b.

Using the law of proportionality,

$$I_{R3} = I_{NT}\frac{R_N}{R_N + R_3} = 0.425 \times \frac{24}{44} = 0.232 \text{ amp}$$

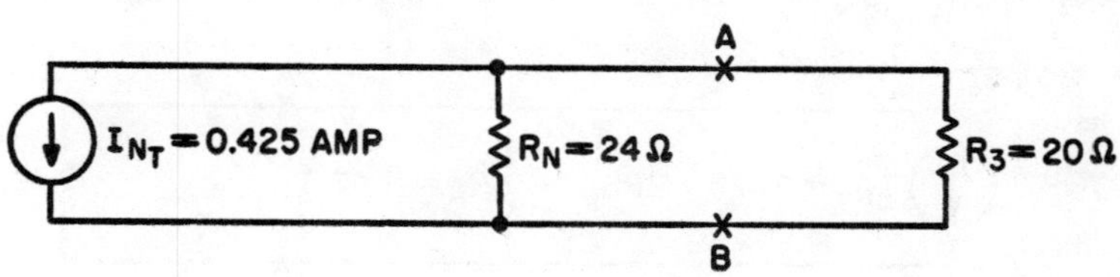

Fig. 3.6c.

3.2 **Find the voltage across R_4 in Fig. 3.7.**

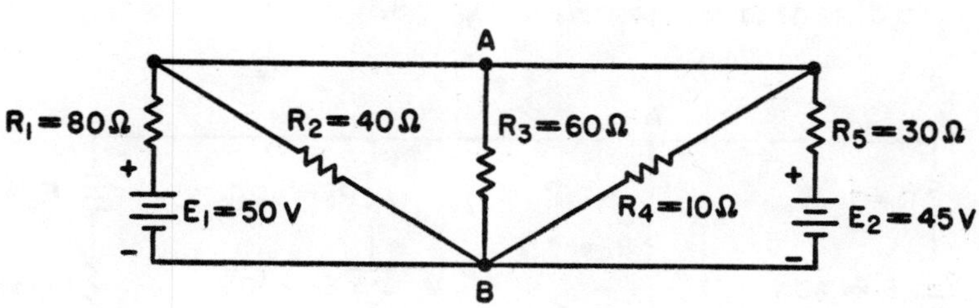

Fig. 3.7.

Answer

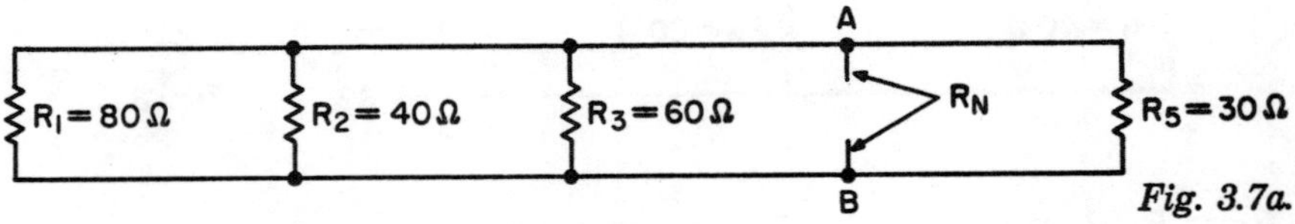

Fig. 3.7a.

$$R_N = \frac{1}{\frac{1}{R_1}+\frac{1}{R_2}+\frac{1}{R_3}+\frac{1}{R_5}} = \frac{1}{\frac{1}{80}+\frac{1}{40}+\frac{1}{60}+\frac{1}{30}} = 11.43 \text{ ohms}$$

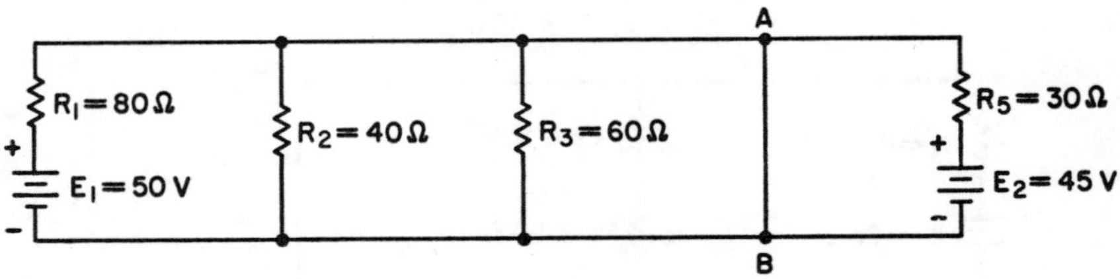

Fig. 3.7b.

In determining I_N (short circuit current), the resistors R_2 and R_3 can be disregarded as they are shorted out.

$$I_{N1} = E_1/R_1 = 50/80 = 0.625 \text{ amp flowing from B to A}$$

$$I_{N2} = E_2/R_5 = 45/30 = 1.5 \text{ amps flowing from B to A}$$

$$I_{NT} = I_{N1} + I_{N2} = 0.625 + 1.5 = 2.125 \text{ amps flowing from B to A}$$

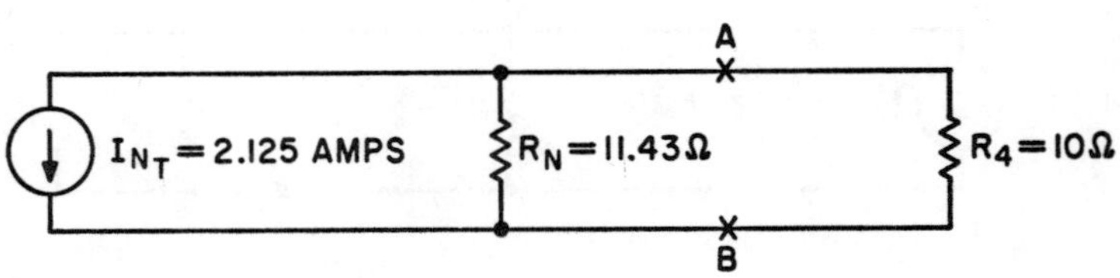

Fig. 3.7c.

R_N and R_4 are in parallel with each other to form R_T. Therefore,

$$E_{R4} = I_{NT} \times R_T = I_{NT}\frac{R_4R_N}{R_4 + R_N} = 2.125 \times \frac{10 \times 11.43}{10 + 11.43} = 11.34 \text{ volts}$$

3.3 Find the voltage across R_4 in Fig. 3.8.

Fig. 3.8.

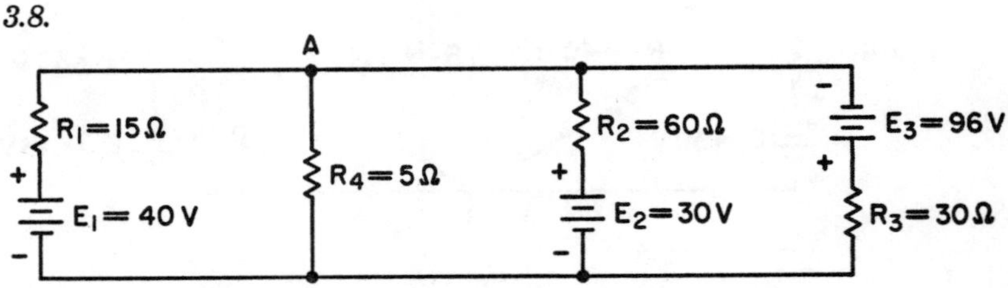

Answer

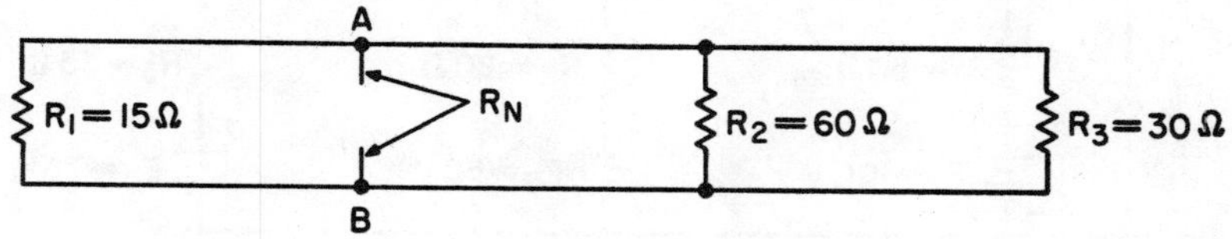

Fig. 3.8a.

$$R_N = \frac{1}{\frac{1}{R_1}+\frac{1}{R_2}+\frac{1}{R_3}} = \frac{1}{\frac{1}{15}+\frac{1}{60}+\frac{1}{30}} = 8.57 \text{ ohms}$$

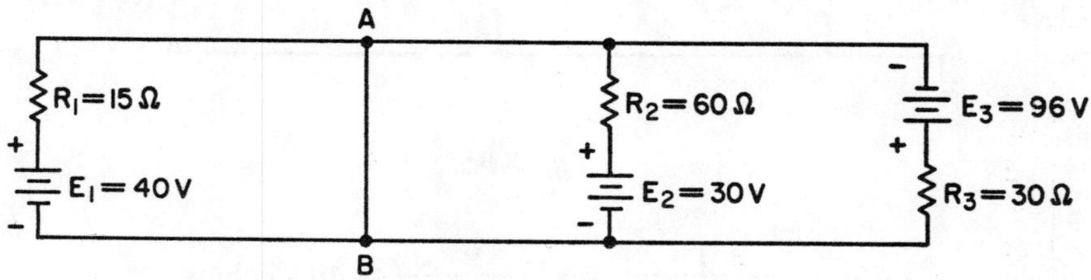

Fig. 3.8b.

$I_{N1} = E_1/R_1 = 40/15 = 2.67$ amps flowing from B to A

$I_{N2} = E_2/R_2 = 30/60 = 0.5$ amp flowing from B to A

$I_{N3} = E_3/R_3 = 96/30 = 3.2$ amps flowing from A to B

$I_{NT} = I_{N3} - (I_{N1} + I_{N2}) = 3.2 - (2.67 + 0.5)$

$= 0.03$ amp flowing from A to B

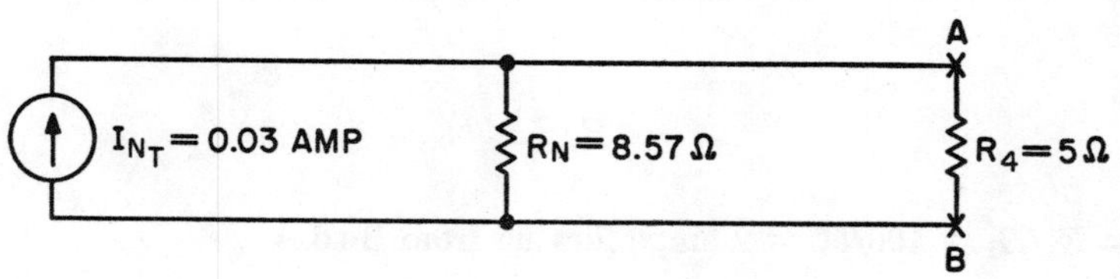

Fig. 3.8c.

R_N and R_4 are in parallel with each other to form R_T. Therefore,

$$E_{R4} = I_{NT} \times R_T = I_{NT}\frac{R_N R_4}{R_N + R_4} = 0.03 \times \frac{8.57 \times 5}{8.57 + 5}$$

$$E_{R4} = 94.8 \text{ millivolts}$$

3.4 Find e_o in Fig. 3.9.

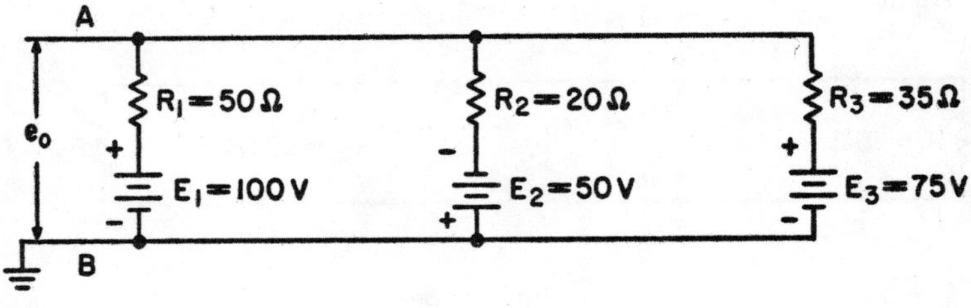

Fig. 3.9.

Answer

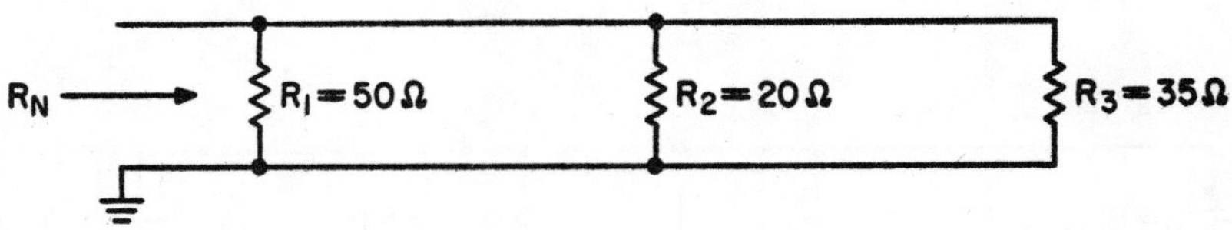

Fig. 3.9a.

$$R_N = \frac{1}{\dfrac{1}{R_1}+\dfrac{1}{R_2}+\dfrac{1}{R_3}} = \frac{1}{\dfrac{1}{50}\quad\dfrac{1}{20}+\dfrac{1}{35}} = 10.15 \text{ ohms}$$

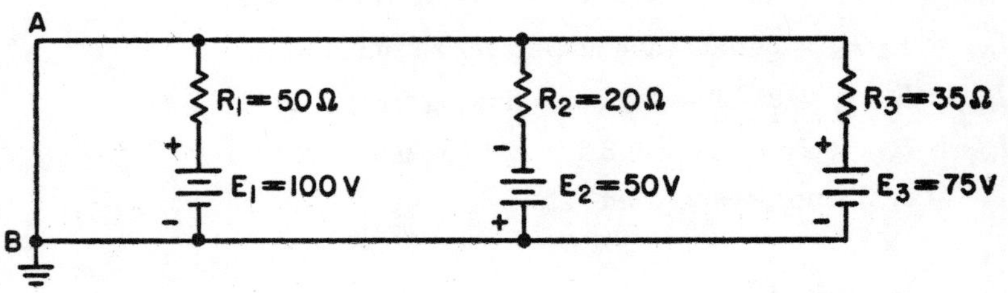

Fig. 3.9b.

$I_{N1} = E_1/R_1 = 100/50 = 2$ amps flowing from B to A

$I_{N2} = E_2/R_2 = 50/20 = 2.5$ amps flowing from A to B

$I_{N3} = E_3/R_3 = 75/35 = 2.14$ amps flowing from B to A

$I_{NT} = I_{N1} + I_{N3} - I_{N2} = 2 + 2.14 - 2.5 = 1.64$ amps flowing from B to A

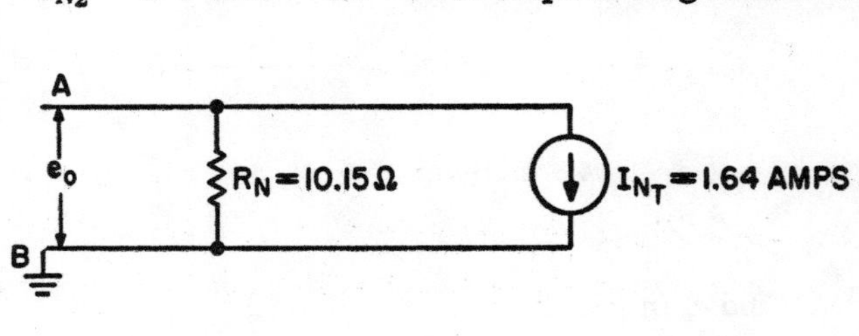

Fig. 3.9c.

$$e_o = I_{NT} \times R_N = 1.64 \times 10.15 = 16.6 \text{ volts}$$

3.5 Find the current flowing through R_5 in Fig. 3.10.

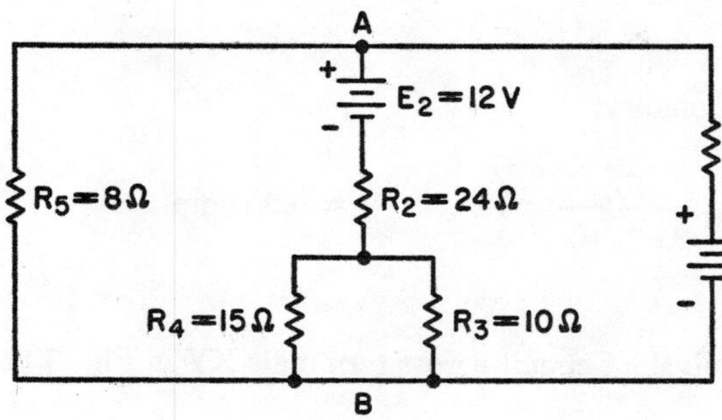

Fig. 3.10.

Answer

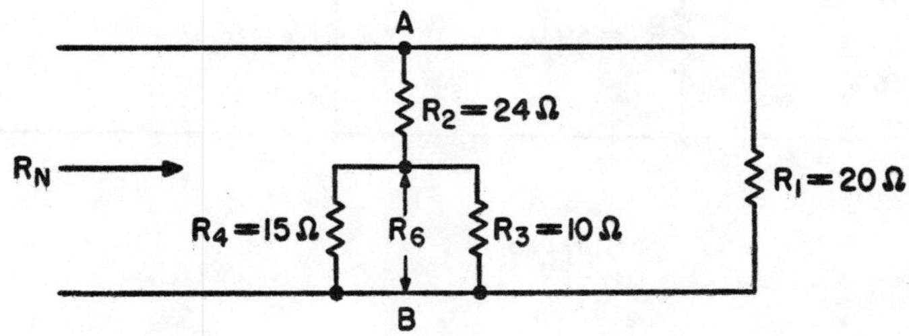

Fig. 3.10a.

Resistors R_3 and R_4 are in parallel with each other and form R_6.

$$R_6 = R_3R_4/(R_3 + R_4) = (10 \times 15)/(10 + 15) = 6 \text{ ohms}$$

$$R_N = \frac{(R_2 + R_6)R_1}{R_2 + R_6 + R_1} = \frac{(24 + 6)20}{24 + 6 + 20} = 12 \text{ ohms}$$

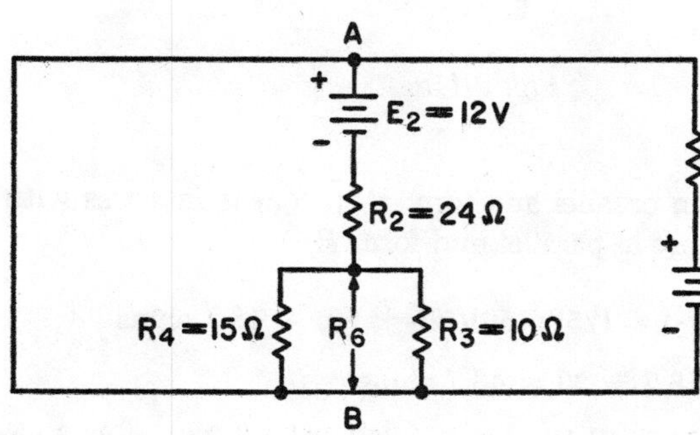

Fig. 3.10b.

$I_{N1} = E_1/R_1 = 25/20 = 1.25$ amps flowing from B to A

$I_{N2} = E_2/(R_2 + R_6) = 12/(24 + 6) = 0.4$ amp flowing from B to A

$I_{NT} = I_{N1} + I_{N2} = 1.25 + 0.4 = 1.65$ amps flowing from B to A

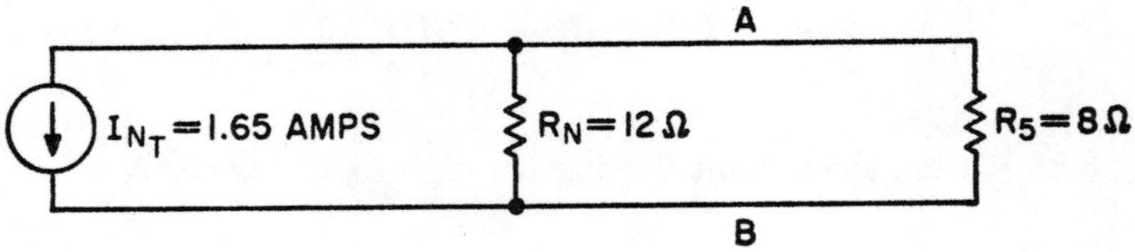

Fig. 3.10c.

Using the law of proportionality,

$$I_{R5} = I_{NT}\frac{R_N}{R_5 + R_N} = 1.65 \times \frac{12}{20} = 0.99 \text{ amp}$$

3.6 Find Norton's equivalent circuit across terminals XY in Fig. 3.11.

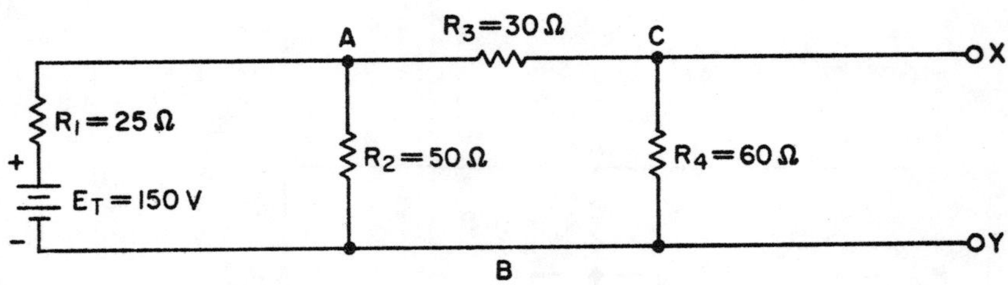

Fig. 3.11.

Answer

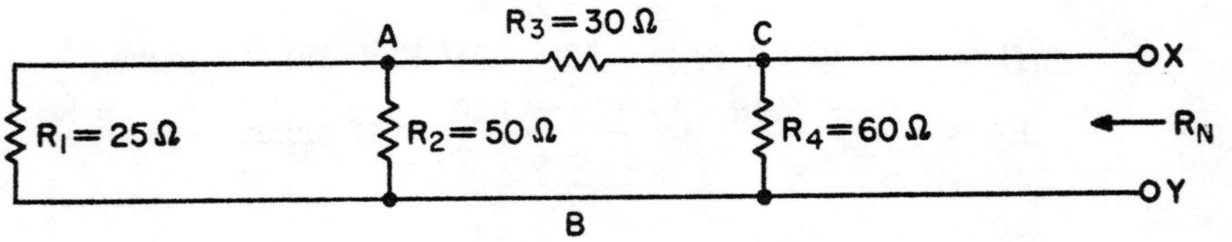

Fig. 3.11a.

Resistors R_1 and R_2 are in parallel and form R_{AB}. R_{AB} is in series with R_3 to form R_{CAB}. R_{CAB} and R_4 are in parallel and form R_N.

$$R_{AB} = R_1R_2/(R_1 + R_2) = (25 \times 50)/(25 + 50) = 16.7 \text{ ohms}$$

$$R_{CAB} = R_{AB} + R_3 = 16.7 + 30 = 46.7 \text{ ohms}$$

$$R_N = (R_{CAB}R_4)/(R_{CAB} + R_4) = (46.7 \times 60)/(46.7 + 60) = 26.3 \text{ ohms}$$

Since R_4 is shorted out, no current will flow through this resistor and I_N will be equal to the current flowing through R_3. Resistors R_2 and R_3 are in parallel with each other and form R_{AB}.

$$R_{AB} = R_2R_3/(R_2 + R_3) = (50 \times 30)/(50 + 30) = 18.75 \text{ ohms}$$

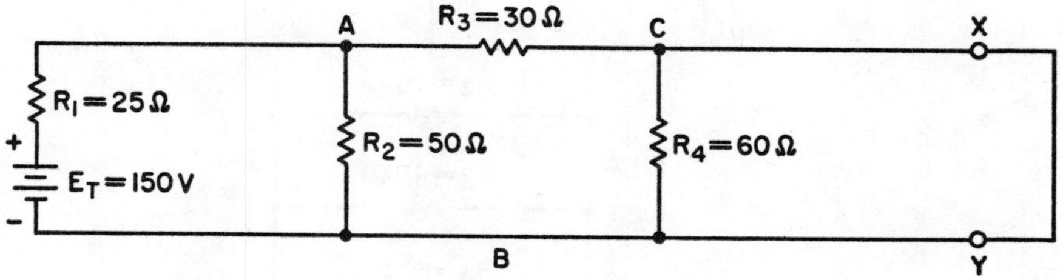

Fig. 3.11b.

You now have a series circuit consisting of E_T, R_1, and R_{AB}. Using the law of proportionality,

$$E_{AB} = E_T \frac{R_{AB}}{R_1 + R_{AB}} = 150 \times \frac{18.75}{43.75} = 64.2 \text{ volts}$$

Since R_2 and R_3 are in parallel and form R_{AB}, $E_{AB} = E_{R_2} = E_{R_3} = 64.2$ volts. Therefore,

$$I_N = E_{R3}/R_3 = 64.2/30 = 2.14 \text{ amps}$$

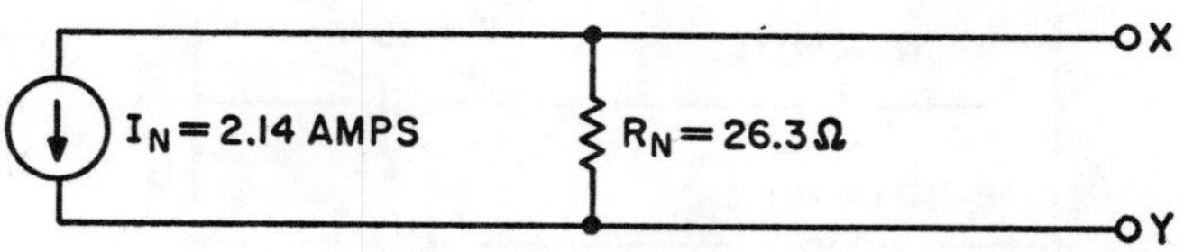

Fig. 3.11c.

3.7 Find the current flowing through R_1 in Fig. 3.12.

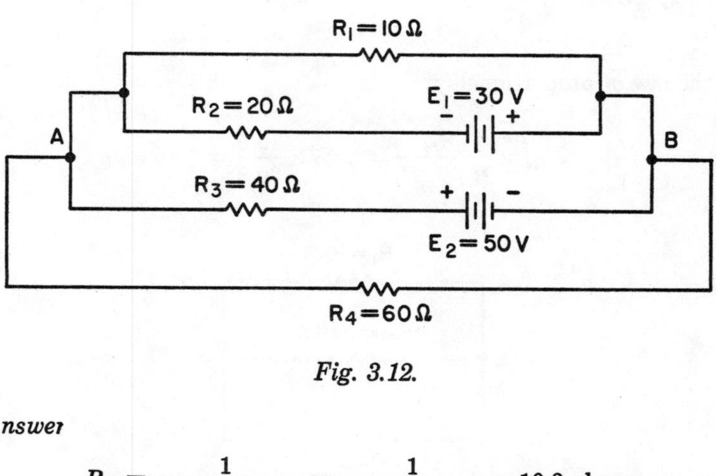

Fig. 3.12.

Answer

$$R_N = \frac{1}{\frac{1}{R_2} + \frac{1}{R_3} + \frac{1}{R_4}} = \frac{1}{\frac{1}{20} + \frac{1}{40} + \frac{1}{60}} = 10.9 \text{ ohms}$$

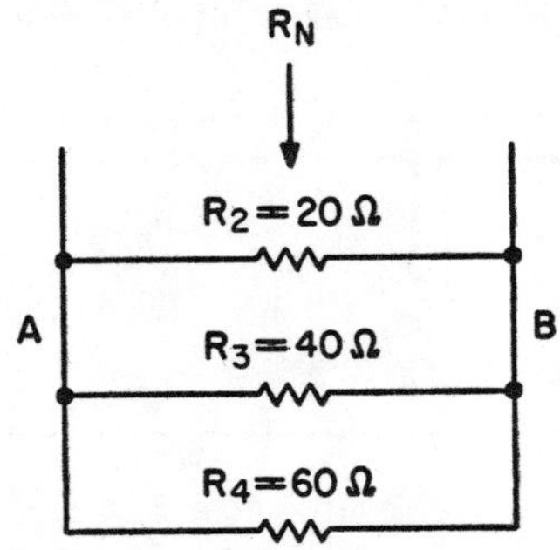

Fig. 3.12a.

In determining I_N (short circuit current), the resistor R_4 can be disregarded as it is shorted out.

$I_{N1} = E_1/R_2 = 30/20 = 1.5$ amps flowing from A to B

$I_{N2} = E_2/R_3 = 50/40 = 1.25$ amps flowing from B to A

$I_{NT} = I_{N1} - I_{N2} = 1.5 - 1.25 = 0.25$ amp flowing from A to B

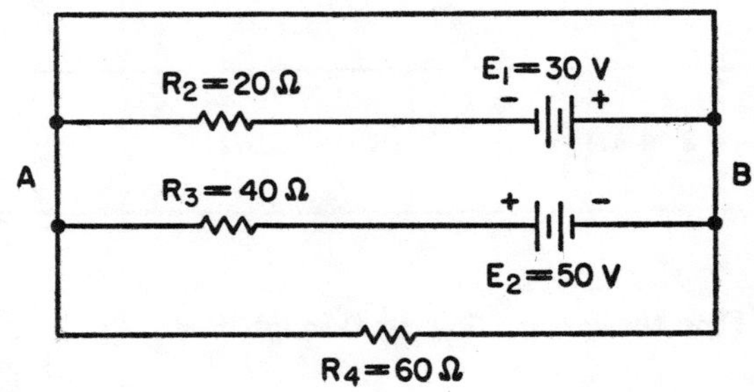

Fig. 3.12b.

Using the law of proportionality,

$$I_{R1} = I_N \times \frac{R_N}{R_N + R_1} = 0.25 \times \frac{10.9}{20.9} = 0.13 \text{ amp}$$

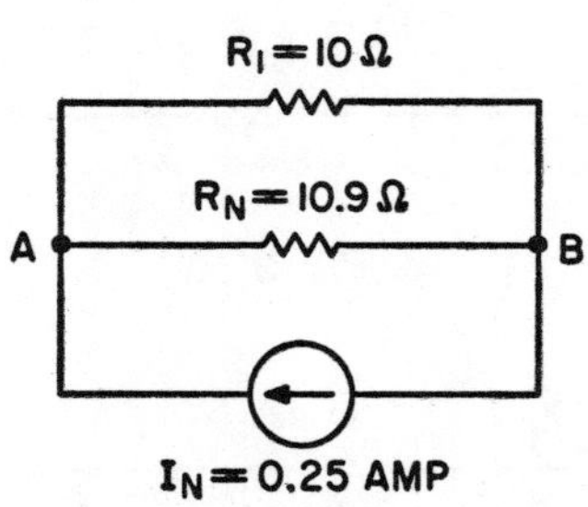

Fig. 3.12c.

3.8 Find the voltage across R_2 in Fig. 3.13.

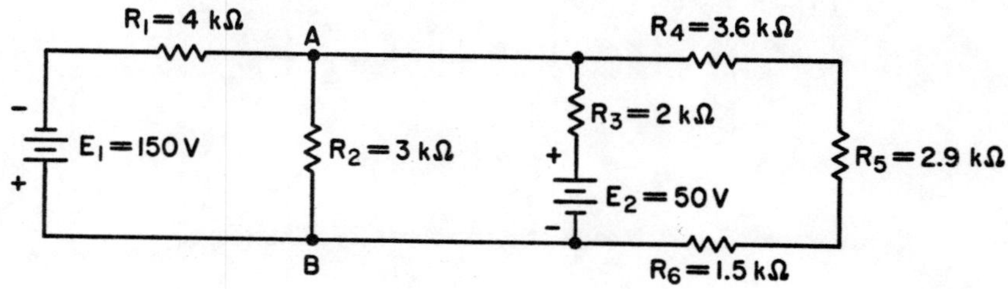

Fig. 3.13.

Answer

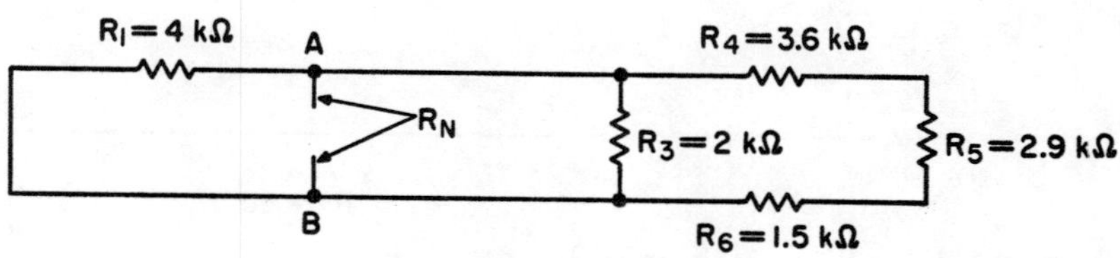

Fig. 3.13a.

$$R_N = \frac{1}{\frac{1}{R_1}+\frac{1}{R_3}+\frac{1}{R_4+R_5+R_6}} = \frac{1}{\frac{1}{4\text{ k}}+\frac{1}{2\text{ k}}+\frac{1}{(3.6+2.9+1.5)\text{ k}}} = 1.143 \text{ kohms}$$

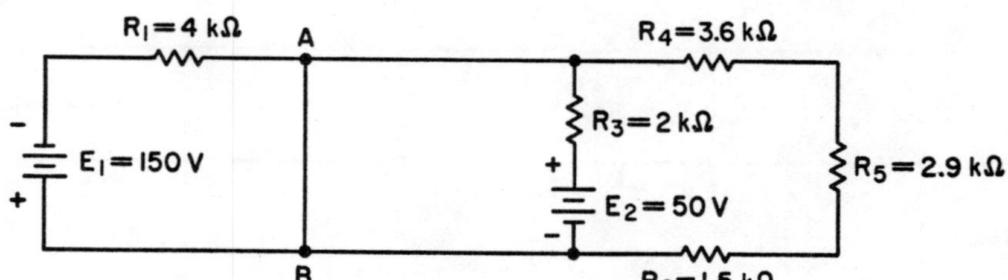

Fig. 3.13b.

In determining I_N (short circuit current), the series combination of R_4, R_5, and R_6 can be disregarded as they are shorted out.

$I_{N1} = E_1/R_1 = 150/4 = 37.5$ ma flowing from A to B

$I_{N2} = E_2/R_3 = 50/2 = 25$ ma flowing from B to A

$I_{NT} = I_{N1} - I_{N2} = 37.5 - 25 = 12.5$ ma flowing from A to B

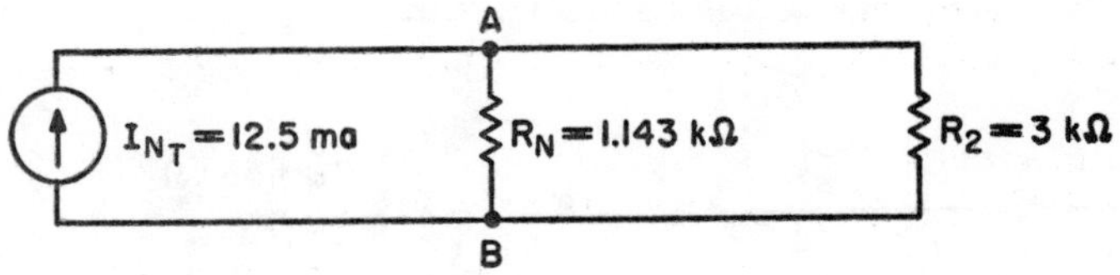

Fig. 3.13c.

R_N and R_2 are in parallel with each other to form R_T. Therefore,

$$E_{R2} = I_{NT} \times R_T = I_{NT} \frac{R_2 R_N}{R_2 + R_N} = 12.5 \times \frac{3 \times 1.143}{3 + 1.143}$$

$$E_{R2} = 10.4 \text{ volts}$$

3.9 Find e_o in Fig. 3.14.

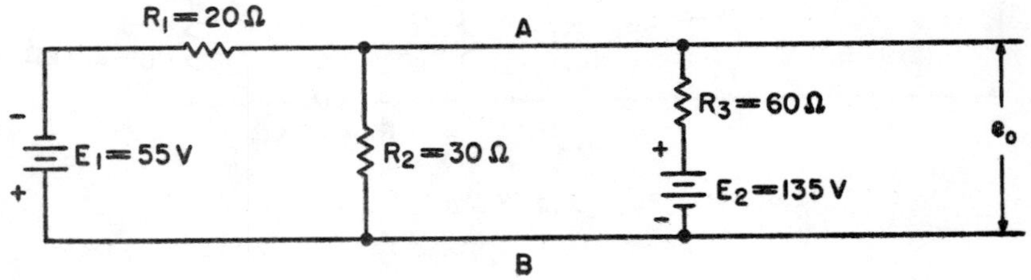

Fig. 3.14.

Answer

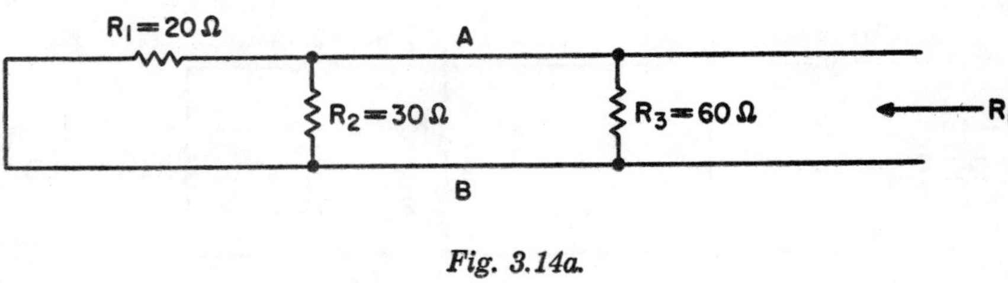

Fig. 3.14a.

$$R_N = \frac{1}{\frac{1}{R_1} + \frac{1}{R_2} + \frac{1}{R_3}} = \frac{1}{\frac{1}{20} + \frac{1}{30} + \frac{1}{60}} = 10 \text{ ohms}$$

In determining I_N (short circuit current), the resistor R_2 can be disregarded since it is shorted out.

$I_{N1} = E_1/R_1 = 55/20 = 2.75$ amps flowing from A to B

$I_{N2} = E_2/R_2 = 135/60 = 2.25$ amps flowing from B to A

$I_{NT} = I_{N1} - I_{N2} = 2.75 - 2.25 = 0.5$ amp flowing from A to B

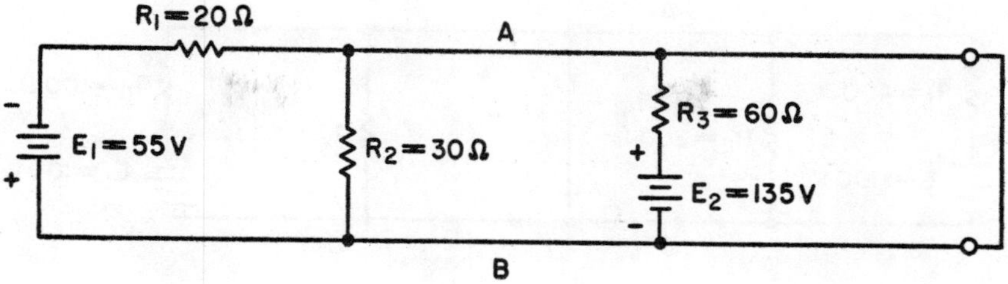

Fig. 3.14b.

$$e_o = I_{NT} \times R_N = 0.5 \times 10 = 5 \text{ volts}$$

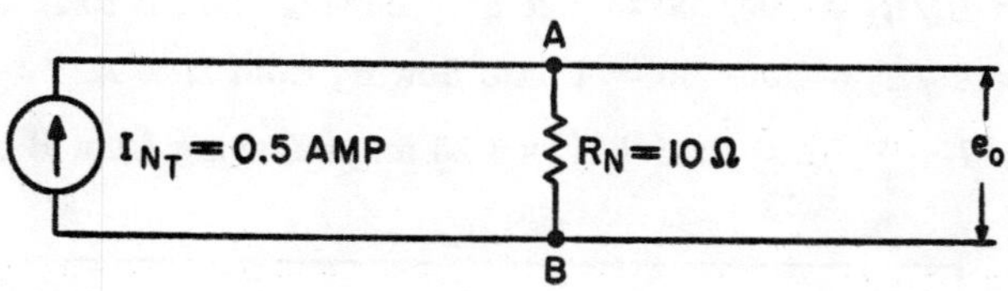

Fig. 3.14c.

3.10 Find the current flowing through R_3 in Fig. 3.15.

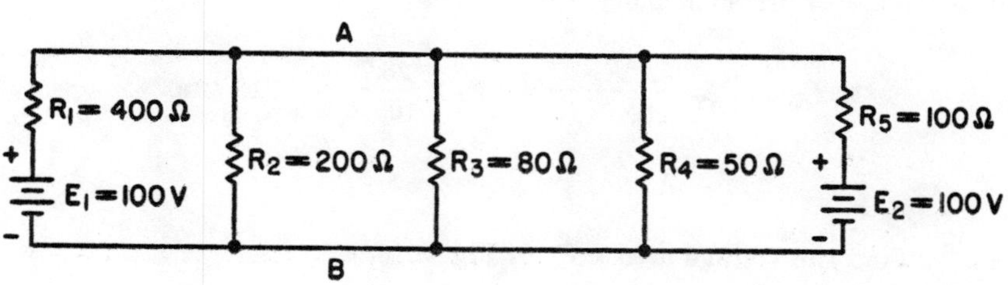

Fig. 3.15.

Answer

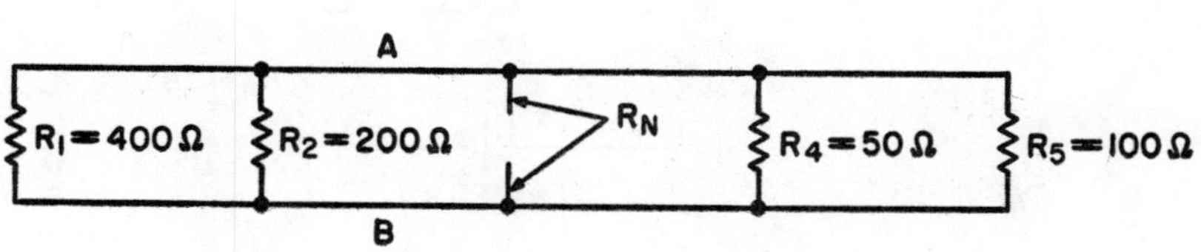

Fig. 3.15a.

$$R_N = \frac{1}{\frac{1}{R_1}+\frac{1}{R_2}+\frac{1}{R_4}+\frac{1}{R_5}} = \frac{1}{\frac{1}{400}+\frac{1}{200}+\frac{1}{50}+\frac{1}{100}} = 26.7 \text{ ohms}$$

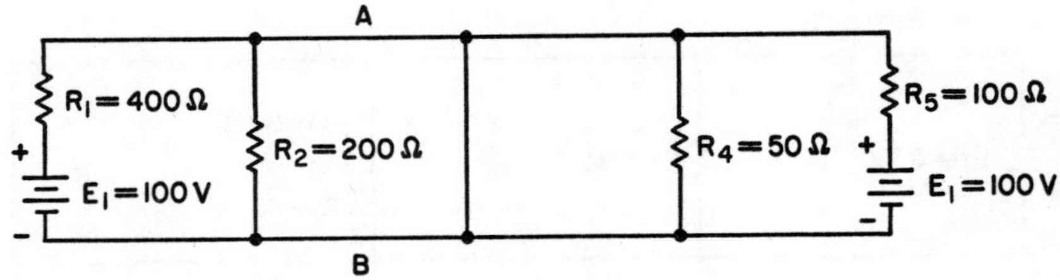

Fig. 3.15b.

In determining I_N (short circuit current), the resistors R_2 and R_4 can be disregarded as they are shorted out.

$$I_{N1} = E_1/R_1 = 100/400 = 0.25 \text{ amp flowing from B to A}$$

$$I_{N2} = E_2/R_5 = 100/100 = 1 \text{ amp flowing from B to A}$$

$$I_{NT} = I_{N1} + I_{N2} = 0.25 + 1 = 1.25 \text{ amps flowing from B to A}$$

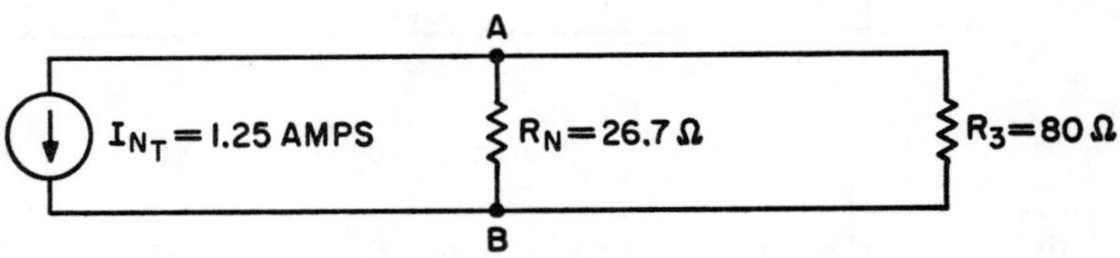

Fig. 3.15c.

Using the law of proportionality,

$$I_{R3} = I_{NT} \frac{R_N}{R_N + R_3} = 1.25 \times \frac{26.7}{106.7} = 0.313 \text{ amp}$$

3.11 Find the voltage across R_5 in Fig. 3.16.

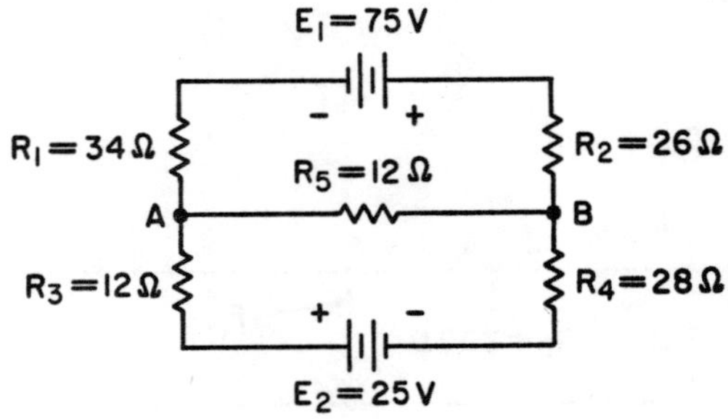

Fig. 3.16.

Answer

$$R_N = \frac{1}{\frac{1}{R_1 + R_2} + \frac{1}{R_3 + R_4}} = \frac{1}{\frac{1}{60} + \frac{1}{40}} = 24 \text{ ohms}$$

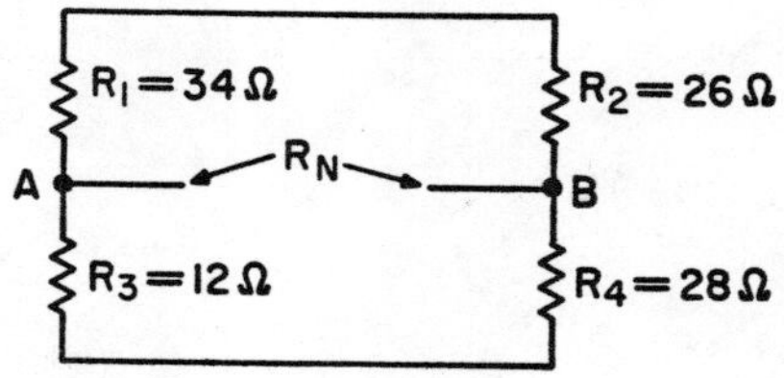

Fig. 3.16a.

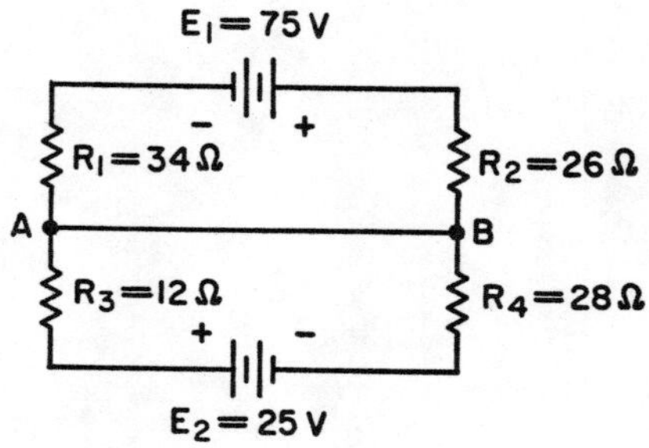

Fig. 3.16b.

$I_{N1} = E_1/(R_1 + R_2) = 75/60 = 1.25$ **amps flowing from A to B**

$I_{N2} = E_2/(R_3 + R_4) = 25/40 = 0.625$ **amp flowing from B to A**

$I_{NT} = I_{N1} - I_{N2} = 1.25 - 0.625 = 0.625$ **amp flowing from A to B**

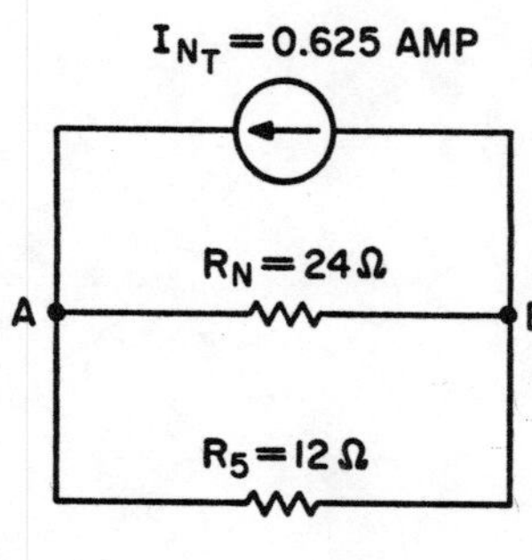

Fig. 3.16c.

R_N and R_5 are in parallel with each other to form R_T. Therefore,

$$E_{R5} = I_{NT}R_T = I_{NT}\frac{R_5R_N}{R_5 + R_N} = 0.625 \times \frac{12 \times 24}{12 + 24}$$

$$E_{R5} = 5 \text{ volts}$$

4

Loop Currents

Chapter 3 of Volume 1 provided solutions to networks with one or more sources of emf contained in the network. The method used was to set up Kirchhoff's law equations and then solve algebraically. The method used in this chapter to solve loop currents is determinants.

Kirchhoff's Laws

1. The sum of all currents flowing *to* any one point in a circuit equals the sum of all currents flowing *from* that point.

$$I_T = I_1 + I_2 + I_3 + I_4$$

or

$$0 = (I_1 + I_2 + I_3 + I_4) - I_T$$

2. Around any closed loop, the algebraic sum of the voltage drops and rises equals zero.

$$E_T = E_1 + E_2 + E_3 + E_4$$

or

$$0 = (E_1 + E_2 + E_3 + E_4) - E_T$$

Loop

Any closed path through the circuit elements of a network.

Loop current

The current that flows in one loop of the network.
To solve problems using loop currents:

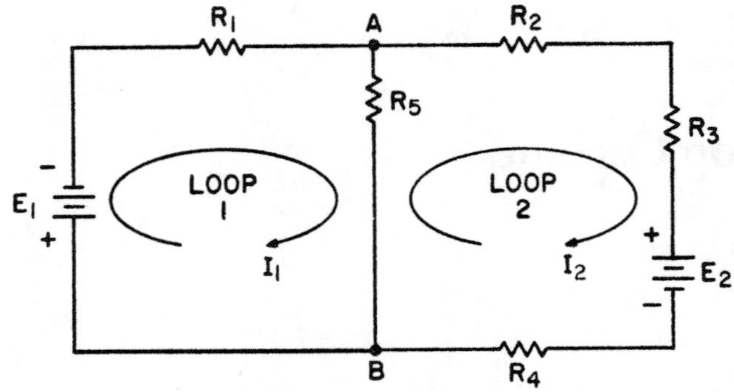

Fig. 4.1.

1. Draw the circuit and assign letters to unknown values: resistance, R; current, I; and voltage, E.

2. Designate each loop as shown in Fig. 4.1. For example: loop 1 and loop 2.

3. Next mark the loop current with a circular arrow as shown in Fig. 4.1 (electron flow is indicated by these loops).

4. You will notice that only one current flows in each loop but that more than one current can flow in a branch. For example, I_1 flows from A to B through R_5 while I_2 flows from B to A through R_5.

5. Now write Kirchhoff's voltage equations for each loop and solve.

Assume a direction for electron current flow for each loop. The end of the resistor where the current leaves is positive with respect to the end where the current enters. To obtain loop equations, travel around a loop (in either direction), noting the voltage drops and rises. These are governed by the assumed direction of the current through the circuit. The equations are then solved for an unknown current by one of the methods for solving simultaneous equations. If a solution for a current yields a negative answer, the negative sign means that the direction for that current was assumed incorrectly. In this case, the correct direction will be opposite to that which was assumed, but the magnitude of the current will still be the same. *Note:* The student should be familiar with determinants in order to solve these equations.

Problems

4.1 Find the current flowing through R_2 in Fig. 4.2.

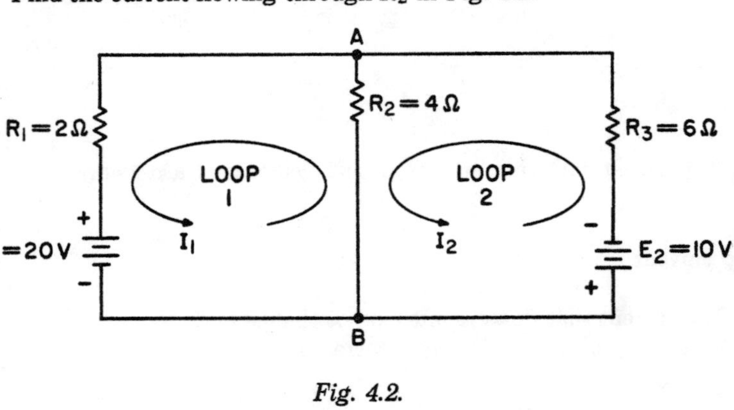

Fig. 4.2.

Answer

$$\text{Loop 1:} \quad 4I_1 - 4I_2 + 2I_1 - 20 = 0$$
$$6I_1 - 4I_2 = 20$$
$$\text{Loop 2:} \quad 6I_2 + 4I_2 - 4I_1 - 10 = 0$$
$$-4I_1 + 10I_2 = 10$$

Using determinants, solve for I_1 and I_2.

$$I_1 = \frac{\begin{vmatrix} 20 & -4 \\ 10 & 10 \end{vmatrix}}{\begin{vmatrix} 6 & -4 \\ -4 & 10 \end{vmatrix}} = \frac{200 + 40}{60 - 16} = \frac{240}{44} = 5.46 \text{ amps}$$

$$I_2 = \frac{\begin{vmatrix} 6 & 20 \\ -4 & 10 \end{vmatrix}}{\begin{vmatrix} 6 & -4 \\ -4 & 10 \end{vmatrix}} = \frac{60 + 80}{44} = \frac{140}{44} = 3.18 \text{ amps}$$

The current flowing through R_2 equals $I_1 - I_2$.

$$I_{R2} = 5.46 - 3.18 = 2.28 \text{ amps flowing from B to A}$$

4.2 Find the current flowing through R_1 in Fig. 4.3.

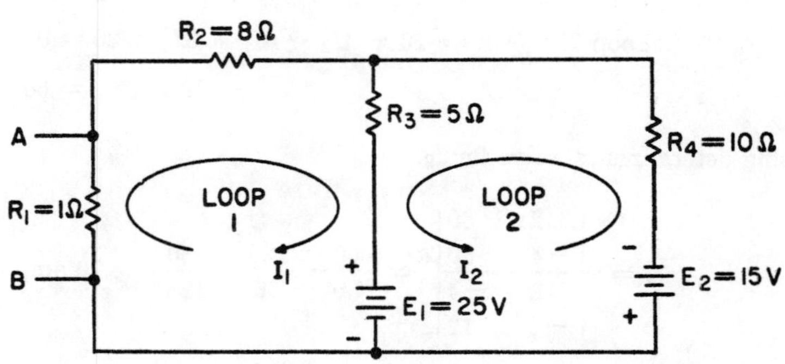

Fig. 4.3.

Answer

$$\text{Loop 1:} \quad I_1 + 8I_1 + 5I_1 + 5I_2 - 25 = 0$$
$$14I_1 + 5I_2 = 25$$
$$\text{Loop 2:} \quad 10I_2 + 5I_2 + 5I_1 - 25 - 15 = 0$$
$$5I_1 + 15I_2 = 40$$

Using determinants, solve for I_1.

$$I_1 = \frac{\begin{vmatrix} 25 & 5 \\ 40 & 15 \end{vmatrix}}{\begin{vmatrix} 14 & 5 \\ 5 & 15 \end{vmatrix}} = \frac{375 - 200}{210 - 25} = \frac{175}{185} = 0.946 \text{ amp}$$

The current flowing through R_1 equals 0.946 amp flowing from B to A.

4.3 Find the current flowing through R_4 in Fig. 4.4.

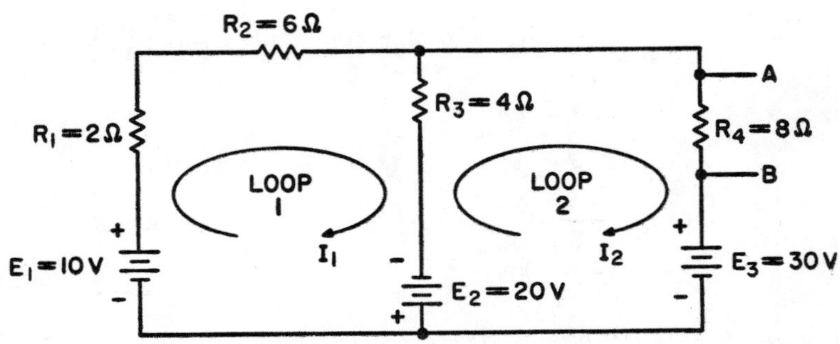

Fig. 4.4.

Answer

Loop 1: $2I_1 + 6I_1 + 4I_1 - 4I_2 + 20 + 10 = 0$

$$12I_1 - 4I_2 = -30$$

Loop 2: $-20 + 4I_2 - 4I_1 + 8I_2 - 30 = 0$

$$-4I_1 + 12I_2 = 50$$

Using determinants, solve for I_2.

$$I_2 = \frac{\begin{vmatrix} 12 & -30 \\ -4 & 50 \end{vmatrix}}{\begin{vmatrix} 12 & -4 \\ -4 & 12 \end{vmatrix}} = \frac{600 - 120}{144 - 16} = \frac{480}{128} = 3.75 \text{ amps}$$

The current flowing through R_4 equals 3.75 amps flowing from A to B.

4.4 Find the current flowing through R_7 in Fig. 4.5.

Answer

Loop 1: $2I_1 + 2I_1 + 2I_2 + 2I_1 - 2I_3 + 2I_1 + 2I_1 - 100 = 0$

$$10I_1 + 2I_2 - 2I_3 = 100$$

$$\text{Loop 2:} \qquad 2I_2 + 2I_2 + 2I_2 + 2I_3 + 2I_2 + 2I_1 = 0$$

$$2I_1 + 8I_2 + 2I_3 = 0$$

$$\text{Loop 3:} \qquad 2I_3 - 2I_1 + 2I_3 + 2I_2 + 2I_3 = 0$$

$$-2I_1 + 2I_2 + 6I_3 = 0$$

Fig. 4.5.

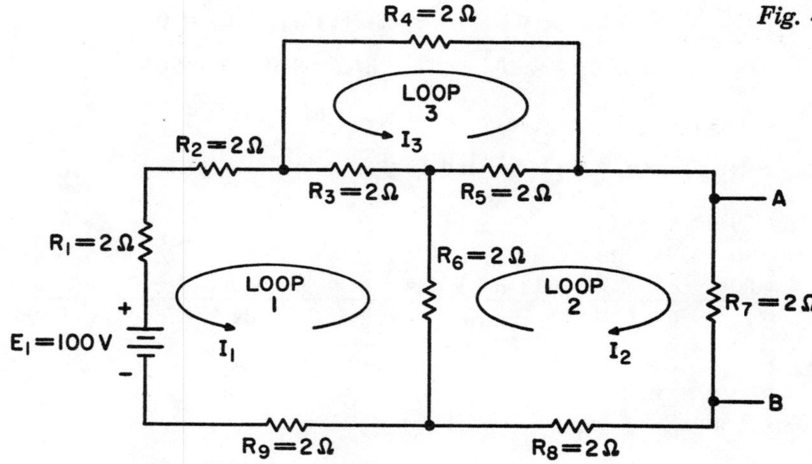

Using determinants, solve for I_2.

$$I_2 = \frac{\begin{vmatrix} 10 & 100 & -2 \\ 2 & 0 & 2 \\ -2 & 0 & 6 \end{vmatrix}}{\begin{vmatrix} 10 & 2 & -2 \\ 2 & 8 & 2 \\ -2 & 2 & 6 \end{vmatrix}} = \frac{0 - 400 + 0 + 0 + 0 - 1{,}200}{480 - 8 - 8 - 32 - 40 - 24} = \frac{-1{,}600}{368}$$

$$I_2 = -4.35 \text{ amps}$$

The current flowing through R_7 is equal to 4.35 amps flowing from B to A, not from A to B as originally assumed.

4.5 Find the voltage drop across R_3 in Fig. 4.6.

Fig. 4.6.

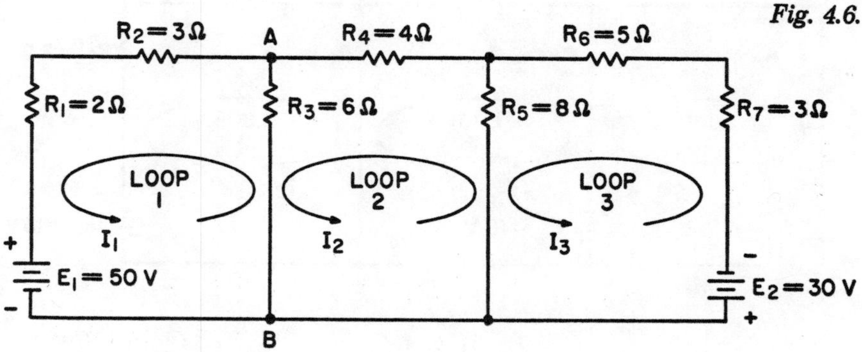

Answer

$$\text{Loop 1:}\quad 6I_1 - 6I_2 + 3I_1 + 2I_1 - 50 = 0$$
$$11I_1 - 6I_2 = 50$$
$$\text{Loop 2:}\quad 8I_2 - 8I_3 + 4I_2 + 6I_2 - 6I_1 = 0$$
$$-6I_1 + 18I_2 - 8I_3 = 0$$
$$\text{Loop 3:}\quad 3I_3 + 5I_3 + 8I_3 - 8I_2 - 30 = 0$$
$$-8I_2 + 16I_3 = 30$$

Using determinants, solve for I_1 and I_2.

$$I_1 = \frac{\begin{vmatrix} 50 & -6 & 0 \\ 0 & 18 & -8 \\ 30 & -8 & 16 \end{vmatrix}}{\begin{vmatrix} 11 & -6 & 0 \\ -6 & 18 & -8 \\ 0 & -8 & 16 \end{vmatrix}} = \frac{14{,}400 + 1{,}440 + 0 + 0 - 3{,}200 + 0}{3{,}170 + 0 + 0 + 0 - 704 - 576} = \frac{12{,}640}{1{,}890}$$
$$= 6.7 \text{ amps}$$

$$I_2 = \frac{\begin{vmatrix} 11 & 50 & 0 \\ -6 & 0 & -8 \\ 0 & 30 & 16 \end{vmatrix}}{\begin{vmatrix} 11 & -6 & 0 \\ -6 & 18 & -8 \\ 0 & -8 & 16 \end{vmatrix}} = \frac{0 + 0 + 0 + 0 + 2{,}640 + 4{,}800}{1{,}890} = \frac{7{,}440}{1{,}890}$$
$$= 3.94 \text{ amps}$$

The current flowing through R_3 is equal to $I_1 - I_2 = 6.7 - 3.94 = 2.76$ amps flowing from B to A. Therefore, the voltage dropped across R_3 is

$$E_{R3} = 2.76 \text{ amps} \times 6 \text{ ohms} = 16.56 \text{ volts}$$

4.6 Find the current flowing through R_5 in Fig. 4.7.

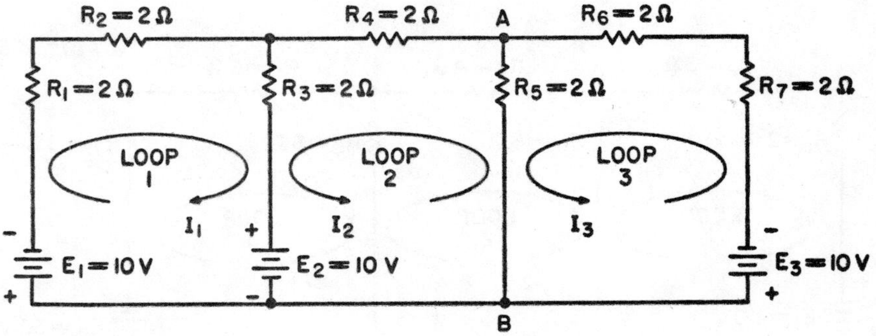

Fig. 4.7.

Answer

$$\text{Loop 1:}\quad 2I_1 + 2I_1 + 2I_1 + 2I_2 - 10 - 10 = 0$$
$$6I_1 + 2I_2 = 20$$
$$\text{Loop 2:}\quad 2I_2 - 2I_3 + 2I_2 + 2I_2 + 2I_1 - 10 = 0$$
$$2I_1 + 6I_2 - 2I_3 = 10$$
$$\text{Loop 3:}\quad 2I_3 + 2I_3 + 2I_3 - 2I_2 - 10 = 0$$
$$-2I_2 + 6I_3 = 10$$

Using determinants, solve for I_2 and I_3.

$$I_2 = \frac{\begin{vmatrix} 6 & 20 & 0 \\ 2 & 10 & -2 \\ 0 & 10 & 6 \end{vmatrix}}{\begin{vmatrix} 6 & 2 & 0 \\ 2 & 6 & -2 \\ 0 & -2 & 6 \end{vmatrix}} = \frac{360 + 0 + 0 + 0 + 120 - 240}{216 + 0 + 0 + 0 - 24 - 24} = \frac{240}{168} = 1.43 \text{ amps}$$

$$I_3 = \frac{\begin{vmatrix} 6 & 2 & 20 \\ 2 & 6 & 10 \\ 0 & -2 & 10 \end{vmatrix}}{\begin{vmatrix} 6 & 2 & 0 \\ 2 & 6 & -2 \\ 0 & -2 & 6 \end{vmatrix}} = \frac{360 + 0 - 80 + 0 + 120 - 40}{168} = \frac{360}{168} = 2.14 \text{ amps}$$

The current flowing through R_5 is equal to $I_3 - I_2 = 2.14 - 1.43 = 0.71$ amp flowing from A to B.

4.7 Find the current flowing through R_2 in Fig. 4.8.

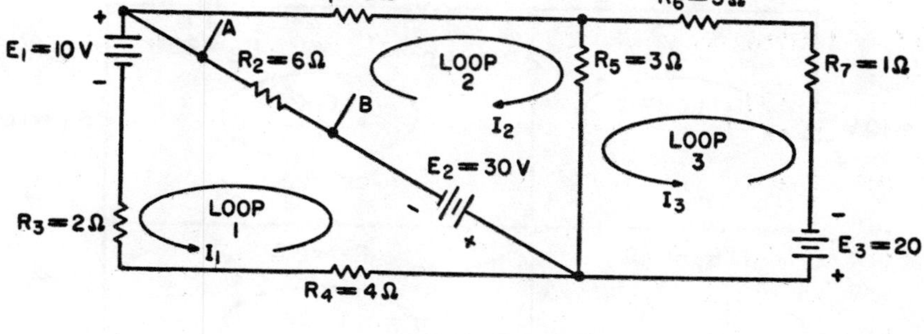

Fig. 4.8.

Answer

$$\text{Loop 1:}\quad 2I_1 + 4I_1 - 30 + 6I_1 + 6I_2 - 10 = 0$$

$$12I_1 + 6I_2 = 40$$

Loop 2: $$6I_2 + 6I_1 + 3I_2 + 3I_2 + 3I_3 - 30 = 0$$

$$6I_1 + 12I_2 + 3I_3 = 30$$

Loop 3: $$I_3 + 5I_3 + 3I_3 + 3I_2 - 20 = 0$$

$$3I_2 + 9I_3 = 20$$

Using determinants, solve for I_1 and I_2.

$$I_1 = \frac{\begin{vmatrix} 40 & 6 & 0 \\ 30 & 12 & 3 \\ 20 & 3 & 9 \end{vmatrix}}{\begin{vmatrix} 12 & 6 & 0 \\ 6 & 12 & 3 \\ 0 & 3 & 9 \end{vmatrix}} = \frac{4{,}320 + 360 + 0 + 0 - 360 - 1{,}620}{1{,}296 + 0 + 0 + 0 - 108 - 324} = \frac{2{,}700}{864}$$

$= 3.13$ amps

$$I_2 = \frac{\begin{vmatrix} 12 & 40 & 0 \\ 6 & 30 & 3 \\ 0 & 20 & 9 \end{vmatrix}}{\begin{vmatrix} 12 & 6 & 0 \\ 6 & 12 & 3 \\ 0 & 3 & 9 \end{vmatrix}} = \frac{3{,}240 + 0 + 0 + 0 - 720 - 2{,}160}{864} = \frac{360}{864}$$

$= 0.417$ amp

The current flowing through R_2 is equal to $I_1 + I_2 = 3.13 + 0.417 = 3.547$ amps flowing from B to A.

4.8 Find the current flowing through R_6 in Fig. 4.9.

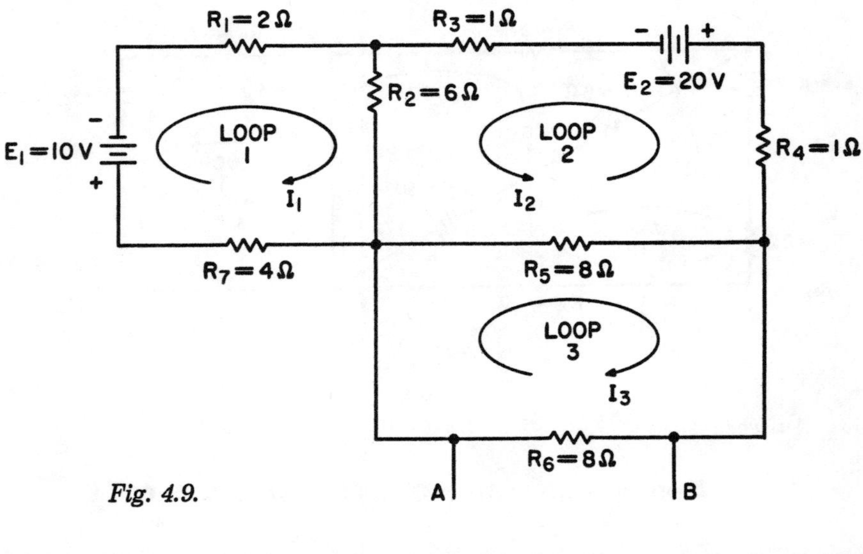

Fig. 4.9.

Answer

Loop 1: $$2I_1 + 6I_1 + 6I_2 + 4I_1 - 10 = 0$$
$$12I_1 + 6I_2 = 10$$

Loop 2: $$I_2 + 6I_2 + 6I_1 + 8I_2 + 8I_3 + I_2 - 20 = 0$$
$$6I_1 + 16I_2 + 8I_3 = 20$$

Loop 3: $$8I_3 + 8I_3 + 8I_2 = 0$$
$$8I_2 + 16I_3 = 0$$

Using determinants, solve for I_3.

$$I_3 = \frac{\begin{vmatrix} 12 & 6 & 10 \\ 6 & 16 & 20 \\ 0 & 8 & 0 \end{vmatrix}}{\begin{vmatrix} 12 & 6 & 0 \\ 6 & 16 & 8 \\ 0 & 8 & 16 \end{vmatrix}} = \frac{0 + 0 + 480 + 0 - 1{,}920 + 0}{3{,}070 + 0 + 0 + 0 - 768 - 576} = \frac{-1{,}440}{1{,}726}$$
$$= -0.833 \text{ amp}$$

The current flowing through R_6 is 0.833 amp flowing from A to B, not B to A as originally assumed.

4.9 Find the current flowing through R_7 in Fig. 4.10.

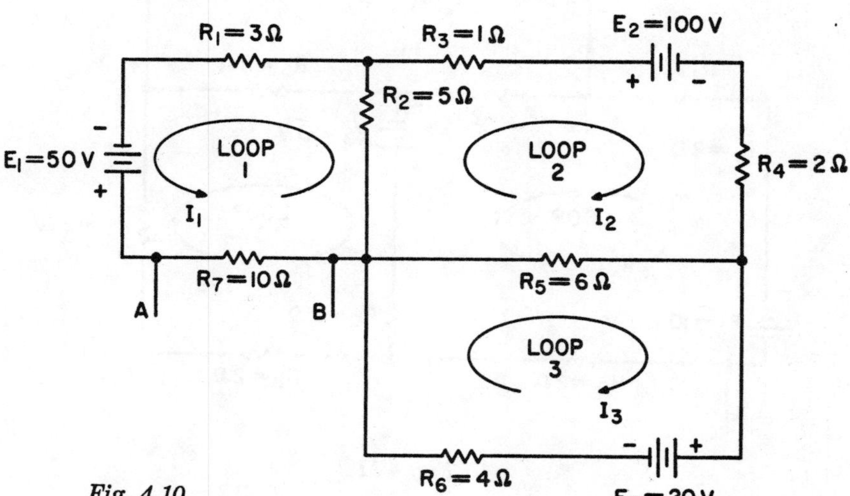

Fig. 4.10.

Answer

Loop 1: $$10I_1 + 5I_1 + 5I_2 + 3I_1 + 50 = 0$$
$$18I_1 + 5I_2 = -50$$

Loop 2: $2I_2 + 6I_2 - 6I_3 + 5I_2 + 5I_1 + I_2 - 100 = 0$

$$5I_1 + 14I_2 - 6I_3 = 100$$

Loop 3: $4I_3 + 6I_3 - 6I_2 - 20 = 0$

$$-6I_2 + 10I_3 = 20$$

Using determinants, solve for I_1.

$$I_1 = \frac{\begin{vmatrix} -50 & 5 & 0 \\ 100 & 14 & -6 \\ 20 & -6 & 10 \end{vmatrix}}{\begin{vmatrix} 18 & 5 & 0 \\ 5 & 14 & -6 \\ 0 & -6 & 10 \end{vmatrix}} = \frac{-7{,}000 - 600 + 0 + 0 + 1{,}800 - 5{,}000}{2{,}520 + 0 + 0 + 0 - 648 - 250} = \frac{-10{,}800}{1{,}622}$$

$I_1 = -6.65$ amps

The current flowing through R_7 equals 6.65 amps flowing from B to A, not from A to B as originally assumed.

4.10 Find the current flowing through R_9 in Fig. 4.11.

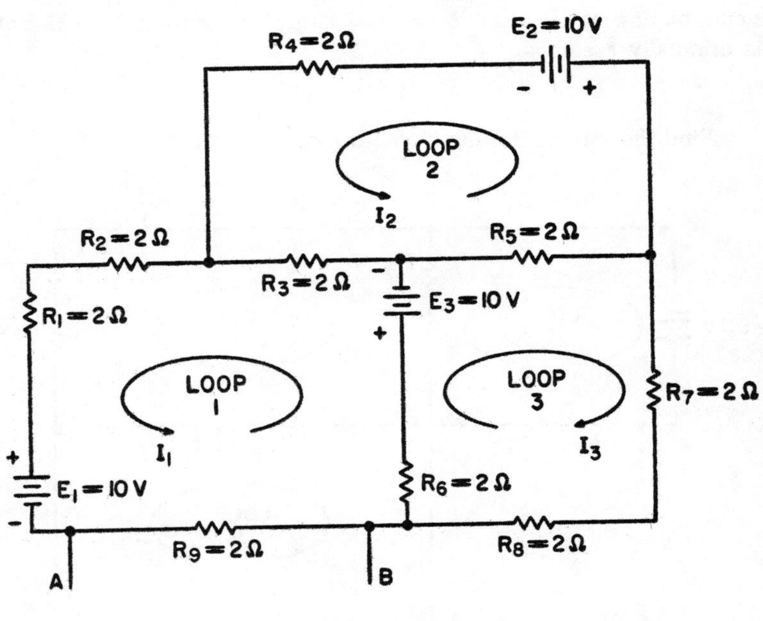

Fig. 4.11.

Answer

Loop 1: $2I_1 + 2I_1 + 2I_3 - 10 + 2I_1 - 2I_2 + 2I_1 + 2I_1 - 10 = 0$

$$10I_1 - 2I_2 + 2I_3 = 20$$

Loop 2: $$2I_2 + 2I_2 - 2I_1 + 2I_2 + 2I_3 - 10 = 0$$
$$-2I_1 + 6I_2 + 2I_3 = 10$$

Loop 3: $$2I_3 + 2I_2 + 2I_3 + 2I_3 + 2I_3 + 2I_1 - 10 = 0$$
$$2I_1 + 2I_2 + 8I_3 = 10$$

Using determinants, solve for I_1.

$$I_1 = \frac{\begin{vmatrix} 20 & -2 & 2 \\ 10 & 6 & 2 \\ 10 & 2 & 8 \end{vmatrix}}{\begin{vmatrix} 10 & -2 & 2 \\ -2 & 6 & 2 \\ 2 & 2 & 8 \end{vmatrix}} = \frac{960 - 40 + 40 - 120 - 80 + 160}{480 - 8 - 8 - 24 - 40 - 32} = \frac{920}{368}$$

$I_1 = 2.5$ amps

The current flowing through R_9 is equal to 2.5 amps flowing from A to B.

4.11 Find the current flowing through R_5 in Fig. 4.12.

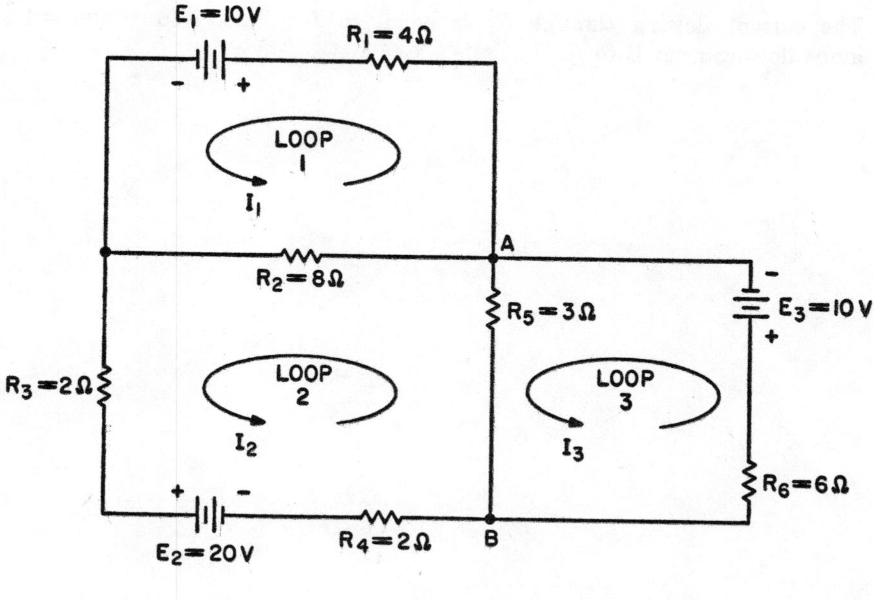

Fig. 4.12.

Answer

Loop 1: $$8I_1 - 8I_2 + 4I_1 - 10 = 0$$
$$12I_1 - 8I_2 = 10$$

Loop 2: $2I_2 + 3I_2 - 3I_3 + 8I_2 - 8I_1 + 2I_2 - 20 = 0$

$$-8I_1 + 15I_2 - 3I_3 = 20$$

Loop 3: $3I_3 - 3I_2 + 6I_3 - 10 = 0$

$$-3I_2 + 9I_3 = 10$$

Using determinants, solve for I_2 and I_3.

$$I_2 = \frac{\begin{vmatrix} 12 & 10 & 0 \\ -8 & 20 & -3 \\ 0 & 10 & 9 \end{vmatrix}}{\begin{vmatrix} 12 & -8 & 0 \\ -8 & 15 & -3 \\ 0 & -3 & 9 \end{vmatrix}} = \frac{2{,}160 + 0 + 0 + 0 + 360 + 720}{1{,}620 + 0 + 0 + 0 - 108 - 576} = \frac{3{,}240}{936} = 3.46 \text{ amps}$$

$$I_3 = \frac{\begin{vmatrix} 12 & -8 & 10 \\ -8 & 15 & 20 \\ 0 & -3 & 10 \end{vmatrix}}{\begin{vmatrix} 12 & -8 & 0 \\ -8 & 15 & -3 \\ 0 & -3 & 9 \end{vmatrix}} = \frac{1{,}800 + 0 + 240 + 0 + 720 - 640}{936} = \frac{2{,}120}{936} = 2.26 \text{ amps}$$

The current flowing through R_5 is equal to $I_2 - I_3 = 3.46 - 2.26 = 1.2$ amps flowing from B to A.

5

Nodal Analysis

Element

A component containing resistance, inductance, capacitance, or current or voltage sources.

Node

A point in a circuit that is common to two or more elements.

Junction

A node that is formed by three or more elements.

Branch

An element or elements connected in series between junctions. If a branch includes one or more ideal sources (constant current or voltage), it is considered active. If no sources are included, the branch is considered passive.

In Chapter 4 problems were solved using loop currents and Kirchhoff's voltage equations. This chapter will solve similar problems but will use branch currents and Kirchhoff's current equations.

In order to write Kirchhoff's current equations, first choose one node in the circuit as a reference point. Then the potential of all other nodes can be expressed with respect to the reference node. Next write Kirchhoff's current equations as shown below. Once these equations have been written, the current values can be obtained.

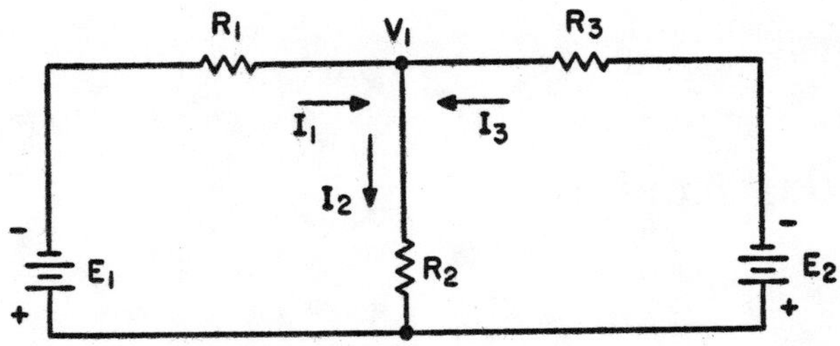

Fig. 5.1.

$$I_1 - I_2 + I_3 = 0$$

$$I_1 = \frac{E_1 - V_1}{R_1}$$

$$I_2 = \frac{V_1}{R_2}$$

$$I_3 = \frac{E_2 - V_1}{R_3}$$

$$\frac{E_1 - V_1}{R_1} - \frac{V_1}{R_2} + \frac{E_2 - V_1}{R_3} = 0$$

Assume a direction for electron current for each branch. The end of the resistor where the current leaves is positive with respect to the end where the current enters. The equations obtained are governed by the assumed direction of the branch current through the circuit. The equations are then solved for an unknown voltage by one of the methods of solving simultaneous equations. If a solution for a voltage yields a negative answer, the negative sign means that the direction for that branch current was assumed incorrectly. The correct direction is opposite to that assumed. The magnitude of the voltage is still the same.

Problems

5.1 Find the current flowing through R_2 in Fig. 5.2.

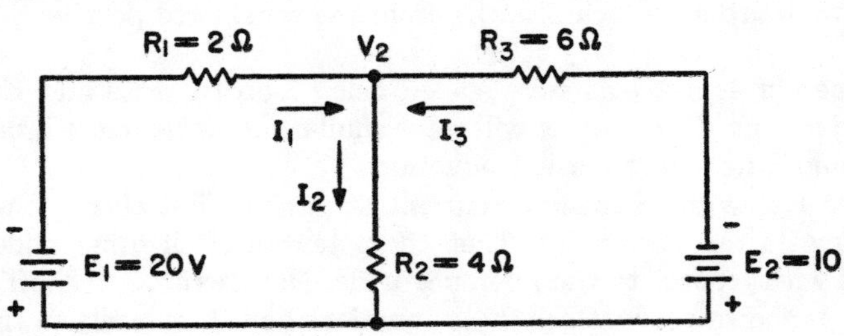

Fig. 5.2.

Answer

$$I_1 = (E_1 - V_2)/R_1 \qquad I_2 = V_2/R_2 \qquad I_3 = (E_2 - V_2)/R_3$$

$$I_2 = I_1 + I_3$$

$$0 = I_1 + I_3 - I_2$$

$$0 = (E_1 - V_2)/(R_1) + (E_2 - V_2)/(R_3) - V_2/R_2$$

$$0 = (20 - V_2)/(2) + (10 - V_2)/(6) - V_2/4$$

The common denominator is 12.

$$0 = 120 - 6V_2 + 20 - 2V_2 - 3V_2$$

$$0 = 140 - 11V_2$$

$$11V_2 = 140$$

$$V_2 = 140/11 = 12.73 \text{ volts}$$

$$I_2 = V_2/R_2 = 12.73/4 = 3.18 \text{ amps}$$

5.2 Find e_o in Fig. 5.3.

Fig. 5.3.

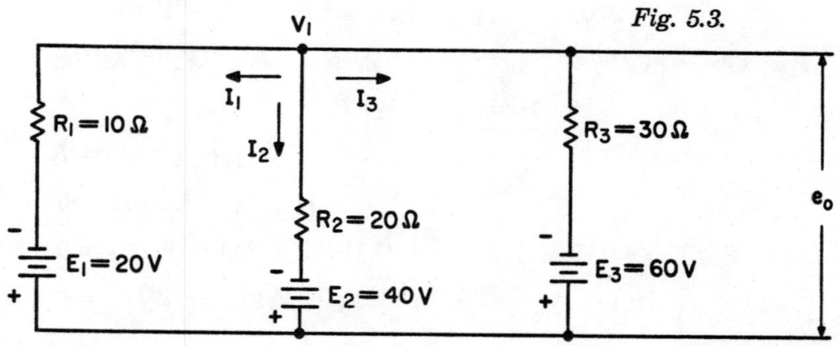

Answer

$$I_1 = (E_1 - V_1)/R_1 \qquad I_2 = (E_2 - V_1)/R_2 \qquad I_3 = (E_3 - V_1)/R_3$$

$$I_1 + I_2 + I_3 = 0$$

$$(E_1 - V_1)/(R_1) + (E_2 - V_1)/(R_2) + (E_3 - V_1)/(R_3) = 0$$

$$(20 - V_1)/(10) + (40 - V_1)/(20) + (60 - V_1)/(30) = 0$$

The common denominator is 60.

$$120 - 6V_1 + 120 - 3V_1 + 120 - 2V_1 = 0$$

$$11V_1 = 360$$

$$V_1 = 360/11 = 32.7 \text{ volts}$$

$$e_o = V_1 = 32.7 \text{ volts}$$

5.3 Find the voltage dropped across R_2 in Fig. 5.4.

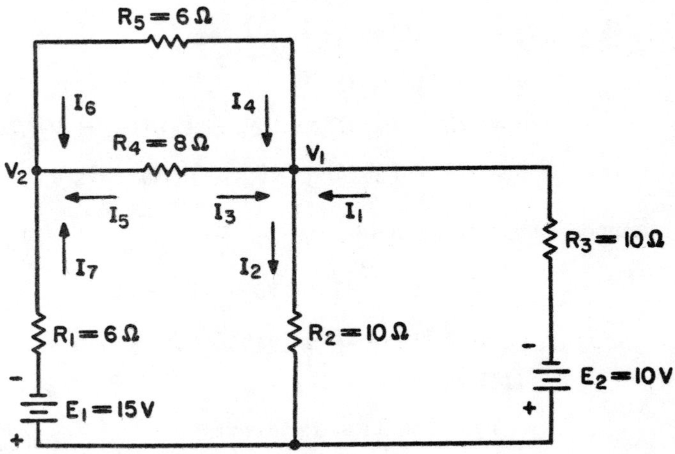

Fig. 5.4.

Answer

$$I_1 = (E_2 - V_1)/R_3 \qquad I_2 = V_1/R_2 \qquad I_3 = (V_2 - V_1)/R_4$$

$$I_4 = (V_2 - V_1)/R_5 \qquad I_5 = (V_1 - V_2)/R_4 \qquad I_6 = (V_1 - V_2)/R_5$$

$$I_7 = (E_1 - V_2)/R_1$$

$$I_1 + I_3 + I_4 = I_2$$

$$I_1 + I_3 + I_4 - I_2 = 0$$

$$(E_2 - V_1)/(R_3) + (V_2 - V_1)/(R_4) + (V_2 - V_1)/(R_5) - V_1/R_2 = 0$$

$$(10 - V_1)/(10) + (V_2 - V_1)/(8) + (V_2 - V_1)/(6) - V_1/10 = 0$$

The common denominator is 120.

$$120 - 12V_1 + 15V_2 - 15V_1 + 20V_2 - 20V_1 - 12V_1 = 0$$

Equation 1:

$$59V_1 - 35V_2 = 120$$

$$I_5 + I_6 + I_7 = 0$$

$$(V_1 - V_2)/(R_4) + (V_1 - V_2)/(R_5) + (E_1 - V_2)/(R_1) = 0$$

$$(V_1 - V_2)/(8) + (V_1 - V_2)/(6) + (15 - V_2)/(6) = 0$$

The common denominator is 24.

$$3V_1 - 3V_2 + 4V_1 - 4V_2 + 60 - 4V_2 = 0$$

Equation 2:

$$-7V_1 + 11V_2 = 60$$

Using determinants, solve for V_1.

$$V_1 = \frac{\begin{vmatrix} 120 & -35 \\ 60 & 11 \end{vmatrix}}{\begin{vmatrix} 59 & -35 \\ -7 & 11 \end{vmatrix}} = \frac{1{,}320 + 2{,}100}{649 - 245} = \frac{3{,}420}{404} = 8.47 \text{ volts}$$

$$E_{R_2} = V_1 = 8.47 \text{ volts}$$

5.4 Find the current flowing through R_3 in Fig. 5.5.

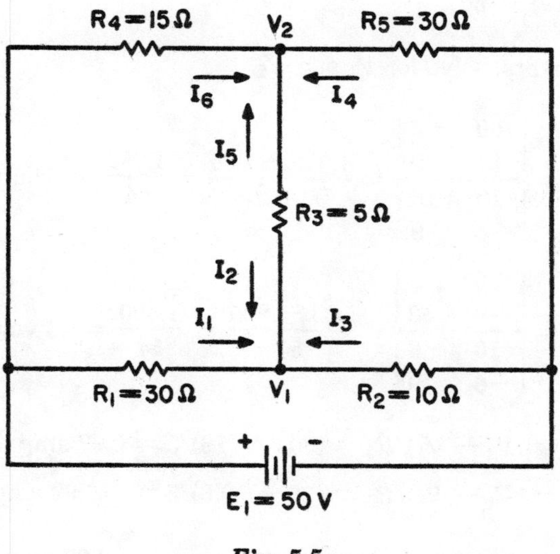

Fig. 5.5.

Answer

$I_1 = -V_1/R_1 \qquad I_2 = (V_2 - V_1)/R_3 \qquad I_3 = (E_1 - V_1)/R_2$

$I_4 = (E_1 - V_2)/R_5 \qquad I_5 = (V_1 - V_2)/R_3 \qquad I_6 = -V_2/R_4$

$$I_1 + I_2 + I_3 = 0$$

$$-V_1/R_1 + (V_2 - V_1)/(R_3) + (E_1 - V_1)/(R_2) = 0$$

$$-V_1/30 + (V_2 - V_1)/(5) + (50 - V_1)/(10) = 0$$

The common denominator is 30.

$$-V_1 + 6V_2 - 6V_1 + 150 - 3V_1 = 0$$

Equation 1:

$$10V_1 - 6V_2 = 150$$

$$I_4 + I_5 + I_6 = 0$$

$$(E_1 - V_2)/(R_5) + (V_1 - V_2)/(R_3) - V_2/R_4 = 0$$

$$(50 - V_2)/(30) + (V_1 - V_2)/5 - V_2/15 = 0$$

The common denominator is 30.

$$50 - V_2 + 6V_1 - 6V_2 - 2V_2 = 0$$

Equation 2:

$$-6V_1 + 9V_2 = 50$$

Using determinants, solve for V_1 and V_2.

$$V_1 = \frac{\begin{vmatrix} 150 & -6 \\ 50 & 9 \end{vmatrix}}{\begin{vmatrix} 10 & -6 \\ -6 & 9 \end{vmatrix}} = \frac{1{,}350 + 300}{90 - 36} = \frac{1{,}650}{54} = 30.6 \text{ volts}$$

$$V_2 = \frac{\begin{vmatrix} 10 & 150 \\ -6 & 50 \end{vmatrix}}{\begin{vmatrix} 10 & -6 \\ -6 & 9 \end{vmatrix}} = \frac{500 + 900}{54} = \frac{1{,}400}{54} = 26 \text{ volts}$$

$$I_2 = (V_2 - V_1)/R_3 = (30.6 - 26)/5 = 0.92 \text{ amp}$$

$$I_5 = (V_1 - V_2)/R_3 = (26 - 30.6)/5 = -0.92 \text{ amp}$$

Therefore, the current flowing through R_3 is equal to 0.92 amp in the direction shown by I_2.

5.5 Find current I_6 flowing through R_5 in Fig. 5.6.

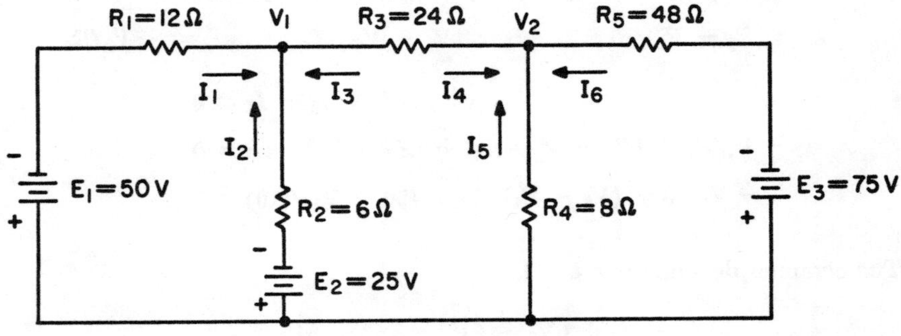

Fig. 5.6.

Answer

$I_1 = (E_1 - V_1)/R_1$ $\quad I_2 = (E_2 - V_1)/R_2$ $\quad I_3 = (V_2 - V_1)/R_3$

$I_4 = (V_1 - V_2)/R_3$ $\quad I_5 = -V_2/R_4$ $\quad I_6 = (E_3 - V_2)/R_5$

$$I_1 + I_2 + I_3 = 0$$

$$(E_1 - V_1)/(R_1) + (E_2 - V_1)/(R_2) + (V_2 - V_1)/(R_3) = 0$$

$$(50 - V_1)/(12) + (25 - V_1)/(6) + (V_2 - V_1)/(24) = 0$$

The common denominator is 24.

$$100 - 2V_1 + 100 - 4V_1 + V_2 - V_1 = 0$$

Equation 1:

$$7V_1 - V_2 = 200$$

$$I_4 + I_5 + I_6 = 0$$

$$(V_1 - V_2)/(R_3) - V_2/(R_4) + (E_3 - V_2)/(R_5) = 0$$

$$(V_1 - V_2)/(24) - V_2/8 + (75 - V_2)/(48) = 0$$

The common denominator is 48.

$$2V_1 - 2V_2 - 6V_2 + 75 - V_2 = 0$$

Equation 2:

$$-2V_1 + 9V_2 = 75$$

Using determinants, solve for V_2.

$$V_2 = \frac{\begin{vmatrix} 7 & 200 \\ -2 & 75 \end{vmatrix}}{\begin{vmatrix} 7 & -1 \\ -2 & 9 \end{vmatrix}} = \frac{525 + 400}{63 - 2} = \frac{925}{61} = 15.16 \text{ volts}$$

$$I_6 = (E_3 - V_2)/R_5 = (75 - 15.16)/48 = 1.246 \text{ amps}$$

5.6 Find the voltage drop across R_1 in Fig. 5.7.

Answer

$I_1 = (E_1 - V_1)/R_1$ $\quad I_2 = -V_1/R_2$ $\quad I_3 = (V_2 - V_1)/R_3$

$I_4 = (V_1 - V_2)/R_3$ $\quad I_5 = -V_2/R_4$ $\quad I_6 = (V_2 - E_2)/R_5$

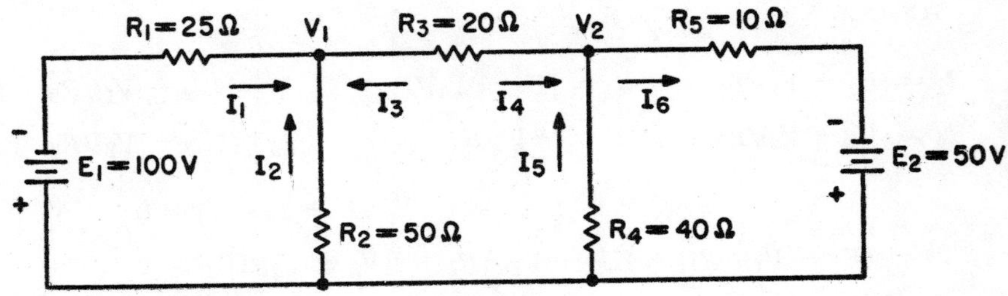

Fig. 5.7.

$$I_1 + I_2 + I_3 = 0$$

$$(E_1 - V_1)/(R_1) - V_1/R_2 + (V_2 - V_1)/(R_3) = 0$$

$$(100 - V_1)/(25) - V_1/50 + (V_2 - V_1)/(20) = 0$$

The common denominator is 100.

$$400 - 4V_1 - 2V_1 + 5V_2 - 5V_1 = 0$$

Equation 1:

$$11V_1 - 5V_2 = 400$$

$$I_4 + I_5 - I_6 = 0$$

$$(V_1 - V_2)/(R_3) - V_2/R_4 - (V_2 - E_2)/(R_5) = 0$$

$$(V_1 - V_2)/(20) - V_2/40 - (V_2 - 50)/(10) = 0$$

The common denominator is 40.

$$2V_1 - 2V_2 - V_2 - 4V_2 + 200 = 0$$

Equation 2:

$$-2V_1 + 7V_2 = 200$$

Using determinants, solve for V_1.

$$V_1 = \frac{\begin{vmatrix} 400 & -5 \\ 200 & 7 \end{vmatrix}}{\begin{vmatrix} 11 & -5 \\ -2 & 7 \end{vmatrix}} = \frac{2{,}800 + 1{,}000}{77 - 10} = \frac{3{,}800}{67} = 56.7 \text{ volts}$$

$$E_{R1} = E_1 - V_1 = 100 - 56.7 = 43.3 \text{ volts}$$

5.7 Find the voltage dropped across R_2 in Fig. 5.8.

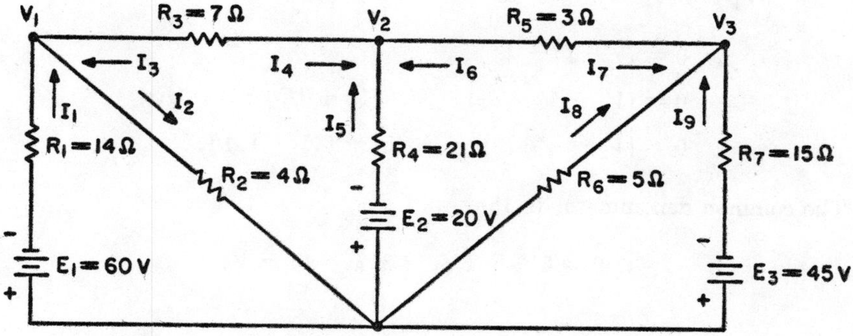

Fig. 5.8.

Answer

$I_1 = (E_1 - V_1)/R_1$ $\quad I_2 = V_1/R_2$ $\quad I_3 = (V_2 - V_1)/R_3$

$I_4 = (V_1 - V_2)/R_3$ $\quad I_5 = (E_2 - V_2)/R_4$ $\quad I_6 = (V_3 - V_2)/R_5$

$I_7 = (V_2 - V_3)/R_5$ $\quad I_8 = -V_3/R_6$ $\quad I_9 = (E_3 - V_3)/R_7$

$$I_2 = I_1 + I_3$$

$$0 = I_1 + I_3 - I_2$$

$$0 = (E_1 - V_1)/(R_1) + (V_2 - V_1)/(R_3) - V_1/R_2$$

$$0 = (60 - V_1)/(14) + (V_2 - V_1)/(7) - V_1/4$$

The common denominator is 28.

$$0 = 120 - 2V_1 + 4V_2 - 4V_1 - 7V_1$$

Equation 1:

$$13V_1 - 4V_2 = 120$$

$$0 = I_4 + I_5 + I_6$$

$$0 = (V_1 - V_2)/(R_3) + (E_2 - V_2)/(R_4) + (V_3 - V_2)/(R_5)$$

$$0 = (V_1 - V_2)/(7) + (20 - V_2)/(21) + (V_3 - V_2)/(3)$$

The common denominator is 21.

$$0 = 3V_1 - 3V_2 + 20 - V_2 + 7V_3 - 7V_2$$

Equation 2:

$$-3V_1 + 11V_2 - 7V_3 = 20$$

$$0 = I_7 + I_8 + I_9$$

$$0 = (V_2 - V_3)/(R_5) - V_3/(R_6) + (E_3 - V_3)/(R_7)$$

$$0 = (V_2 - V_3)/(3) - V_3/(5) + (45 - V_3)/(15)$$

The common denominator is 15.

$$0 = 5V_2 - 5V_3 - 3V_3 + 45 - V_3$$

Equation 3:

$$-5V_2 + 9V_3 = 45$$

Using determinants, solve for V_1.

$$V_1 = \frac{\begin{vmatrix} 120 & -4 & 0 \\ 20 & 11 & -7 \\ 45 & -5 & 9 \end{vmatrix}}{\begin{vmatrix} 13 & -4 & 0 \\ -3 & 11 & -7 \\ 0 & -5 & 9 \end{vmatrix}} = \frac{11{,}880 + 1{,}260 + 0 + 0 - 4{,}200 + 720}{1{,}287 + 0 + 0 + 0 - 455 - 108}$$

$$V_1 = 9{,}660/724 = 13.34 \text{ volts}$$

$$E_{R2} = V_1 = 13.34 \text{ volts}$$

5.8 Find the current flowing through R_2 in Fig. 5.9.

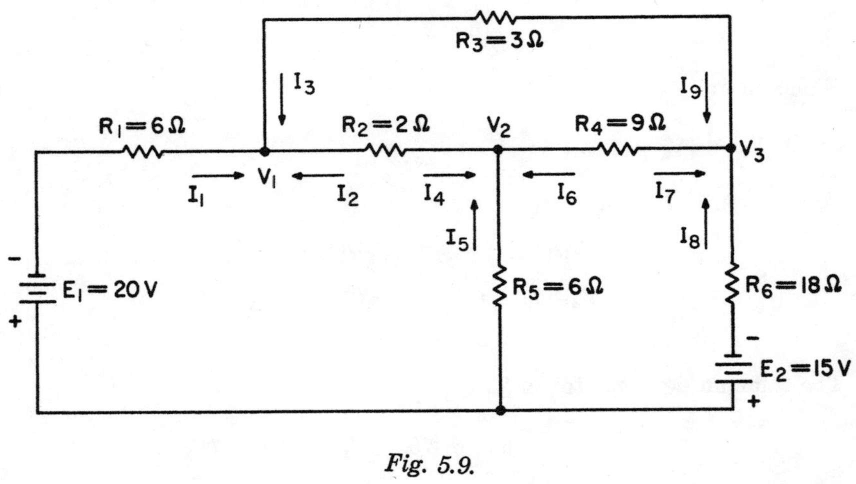

Fig. 5.9.

Answer

$I_1 = (E_1 - V_1)/R_1$ $\qquad I_2 = (V_2 - V_1)/R_2$ $\qquad I_3 = (V_3 - V_1)/R_3$

$I_4 = (V_1 - V_2)/R_2$ $\qquad I_5 = -V_2/R_5$ $\qquad I_6 = (V_3 - V_2)/R_4$

$I_7 = (V_2 - V_3)/R_4$ $\qquad I_8 = (E_2 - V_3)/R_6$ $\qquad I_9 = (V_1 - V_3)/R_3$

$$I_1 + I_2 + I_3 = 0$$

$$(E_1 - V_1)/(R_1) + (V_2 - V_1)/(R_2) + (V_3 - V_1)/(R_3) = 0$$

$$(20 - V_1)/(6) + (V_2 - V_1)/(2) + (V_3 - V_1)/(3) = 0$$

The common denominator is 6.

$$20 - V_1 + 3V_2 - 3V_1 + 2V_3 - 2V_1 = 0$$

Equation 1:

$$-6V_1 + 3V_2 + 2V_3 = -20$$

$$I_4 + I_5 + I_6 = 0$$

$$(V_1 - V_2)/(R_2) - V_2/R_5 + (V_3 - V_2)/(R_4) = 0$$

$$(V_1 - V_2)/(2) - V_2/6 + (V_3 - V_2)/(9) = 0$$

The common denominator is 18.

$$9V_1 - 9V_2 - 3V_2 + 2V_3 - 2V_2 = 0$$

Equation 2:

$$9V_1 - 14V_2 + 2V_3 = 0$$

$$I_7 + I_8 + I_9 = 0$$

$$(V_2 - V_3)/(R_4) + (E_2 - V_3)/(R_6) + (V_1 - V_3)/(R_3) = 0$$

$$(V_2 - V_3)/(9) + (15 - V_3)/(18) + (V_1 - V_3)/(3) = 0$$

The common denominator is 18.

$$2V_2 - 2V_3 + 15 - V_3 + 6V_1 - 6V_3 = 0$$

Equation 3:

$$6V_1 + 2V_2 - 9V_3 = -15$$

Using determinants, solve for V_1 and V_2.

$$V_1 = \frac{\begin{vmatrix} -20 & 3 & 2 \\ 0 & -14 & 2 \\ -15 & 2 & -9 \end{vmatrix}}{\begin{vmatrix} -6 & 3 & 2 \\ 9 & -14 & 2 \\ 6 & 2 & -9 \end{vmatrix}} = \frac{-2{,}520 - 90 + 0 - 420 + 80 + 0}{-756 + 36 + 36 + 168 + 24 + 243}$$

$$V_1 = -2{,}950/-249 = 11.84 \text{ volts}$$

$$V_2 = \frac{\begin{vmatrix} -6 & -20 & 2 \\ 9 & 0 & 2 \\ 6 & -15 & -9 \end{vmatrix}}{\begin{vmatrix} -6 & 3 & 2 \\ 9 & -14 & 2 \\ 6 & 2 & -9 \end{vmatrix}} = \frac{0 - 240 - 270 + 0 - 180 - 1{,}620}{-249}$$

$$V_2 = -2{,}310/-249 = 9.28 \text{ volts}$$

$$I_2 = (V_2 - V_1)/R_2 = (9.28 - 11.84)/2 = -1.28 \text{ amps}$$

$$I_4 = (V_1 - V_2)/R_2 = (11.84 - 9.28)/2 = 1.28 \text{ amps}$$

The current flowing through R_2 is 1.28 amps as shown by I_4.

5.9 Find the voltage dropped across R_7 in Fig. 5.10.

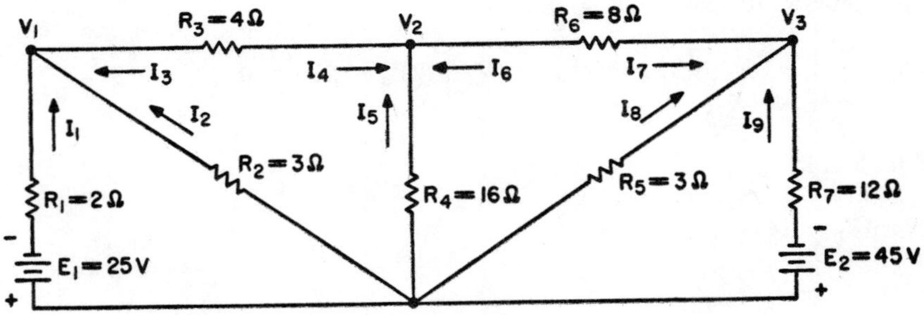

Fig. 5.10.

Answer

$I_1 = (E_1 - V_1)/R_1 \qquad I_2 = -V_1/R_2 \qquad I_3 = (V_2 - V_1)/R_3$

$I_4 = (V_1 - V_2)/R_3 \qquad I_5 = -V_2/R_4 \qquad I_6 = (V_3 - V_2)/R_6$

$I_7 = (V_2 - V_3)/R_6 \qquad I_8 = -V_3/R_5 \qquad I_9 = (E_2 - V_3)/R_7$

$$I_1 + I_2 + I_3 = 0$$

$$(E_1 - V_1)/(R_1) - V_1/R_2 + (V_2 - V_1)/(R_3) = 0$$

$$(25 - V_1)/(2) - V_1/3 + (V_2 - V_1)/(4) = 0$$

The common denominator is 12.

$$150 - 6V_1 - 4V_1 + 3V_2 - 3V_1 = 0$$

Equation 1:

$$-13V_1 + 3V_2 = -150$$

$$I_4 + I_5 + I_6 = 0$$
$$(V_1 - V_2)/(R_3) - V_2/R_4 + (V_3 - V_2)/(R_6) = 0$$
$$(V_1 - V_2)/(4) - V_2/16 + (V_3 - V_2)/(8) = 0$$

The common denominator is 16.

$$4V_1 - 4V_2 - V_2 + 2V_3 - 2V_2 = 0$$

Equation 2:

$$4V_1 - 7V_2 + 2V_3 = 0$$

$$I_7 + I_8 + I_9 = 0$$
$$(V_2 - V_3)/(R_6) - V_3/R_5 + (E_2 - V_3)/(R_7) = 0$$
$$(V_2 - V_3)/(8) - V_3/3 + (45 - V_3)/(12) = 0$$

The common denominator is 24.

$$3V_2 - 3V_3 - 8V_3 + 90 - 2V_3 = 0$$

Equation 3:

$$3V_2 - 13V_3 = -90$$

Using determinants, solve for V_3.

$$V_3 = \frac{\begin{vmatrix} -13 & 3 & -150 \\ 4 & -7 & 0 \\ 0 & 3 & -90 \end{vmatrix}}{\begin{vmatrix} -13 & 3 & 0 \\ 4 & -7 & 2 \\ 0 & 3 & -13 \end{vmatrix}} = \frac{-8{,}190 + 0 - 1{,}800 + 0 + 0 + 1{,}080}{-1{,}183 + 0 + 0 + 0 + 78 + 156}$$

$$V_3 = -8{,}910/-949 = 9.4 \text{ volts}$$

$$E_{R7} = E_2 - V_3 = 45 - 9.4 = 35.6 \text{ volts}$$

5.10 Find the current flowing through R_3 in Fig. 5.11.

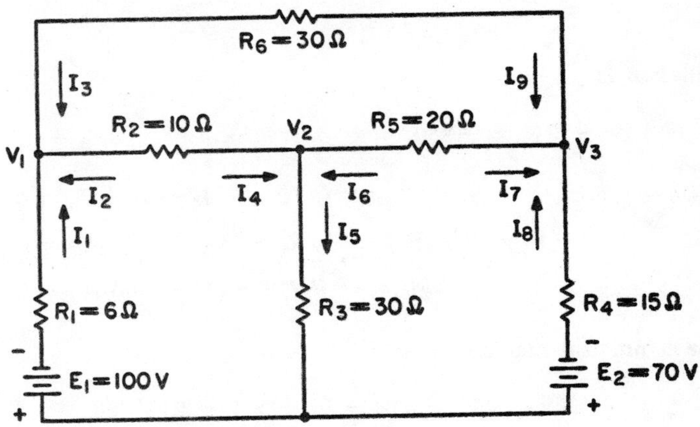

Fig. 5.11.

Answer

$I_1 = (E_1 - V_1)/R_1$ $\quad I_2 = (V_2 - V_1)/R_2$ $\quad I_3 = (V_3 - V_1)/R_6$

$I_4 = (V_1 - V_2)/R_2$ $\quad I_5 = V_2/R_3$ $\quad I_6 = (V_3 - V_2)/R_5$

$I_7 = (V_2 - V_3)/R_5$ $\quad I_8 = (E_2 - V_3)/R_4$ $\quad I_9 = (V_1 - V_3)/R_6$

$$I_1 + I_2 + I_3 = 0$$

$$(E_1 - V_1)/(R_1) + (V_2 - V_1)/(R_2) + (V_3 - V_1)/(R_6) = 0$$

$$(100 - V_1)/(6) + (V_2 - V_1)/(10) + (V_3 - V_1)/(30) = 0$$

The common denominator is 30.

$$500 - 5V_1 + 3V_2 - 3V_1 + V_3 - V_1 = 0$$

Equation 1:

$$-9V_1 + 3V_2 + V_3 = -500$$

$$I_4 + I_6 = I_5$$

$$I_4 + I_6 - I_5 = 0$$

$$(V_1 - V_2)/(R_2) + (V_3 - V_2)/(R_5) + V_2/R_3 = 0$$

$$(V_1 - V_2)/(10) + (V_3 - V_2)/(20) + V_2/(30) = 0$$

The common denominator is 60.

$$6V - 6V_2 + 3V_3 - 3V_2 + 2V_2 = 0$$

Equation 2:

$$6V_1 - 7V_2 + 3V_3 = 0$$

$$I_7 + I_8 + I_9 = 0$$

$$(V_2 - V_3)/(R_5) + (E_2 - V_3)/(R_4) + (V_1 - V_3)/(R_6) = 0$$

$$(V_2 - V_3)/(20) + (70 - V_3)/(15) + (V_1 - V_3)/(30) = 0$$

The common denominator is 60.

$$3V_2 - 3V_3 + 280 - 4V_3 + 2V_1 - 2V_3 = 0$$

Equation 3:

$$2V_1 + 3V_2 - 9V_3 = -280$$

Using determinants, solve for V_2.

$$V_2 = \frac{\begin{vmatrix} -9 & -500 & 1 \\ 6 & 0 & 3 \\ 2 & -280 & -9 \end{vmatrix}}{\begin{vmatrix} -9 & 3 & 1 \\ 6 & -7 & 3 \\ 2 & 3 & -9 \end{vmatrix}} = \frac{0 - 3{,}000 - 1{,}680 + 0 - 7{,}560 - 27{,}000}{-567 + 18 + 18 + 14 + 81 + 162}$$

$$V_2 = -39{,}240/-274 = 143 \text{ volts}$$

$$I_5 = V_2/R_3 = 143/30 = 4.77 \text{ amps}$$

5.11 Find the current flowing through R_1 in Fig. 5.12.

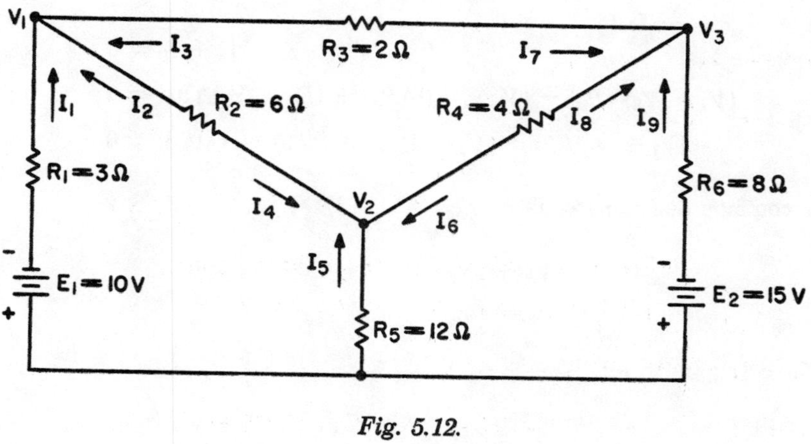

Fig. 5.12.

Answer

$I_1 = (E_1 - V_1)/R_1$ $I_2 = (V_2 - V_1)/R_2$ $I_3 = (V_3 - V_1)/R_3$

$I_4 = (V_1 - V_2)/R_2$ $I_5 = -V_2/R_5$ $I_6 = (V_3 - V_2)/R_4$

$I_7 = (V_1 - V_3)/R_3$ $I_8 = (V_2 - V_3)/R_4$ $I_9 = (E_2 - V_3)/R_6$

$$I_1 + I_2 + I_3 = 0$$
$$(E_1 - V_1)/(R_1) + (V_2 - V_1)/(R_2) + (V_3 - V_1)/(R_3) = 0$$
$$(10 - V_1)/(3) + (V_2 - V_1)/(6) + (V_3 - V_1)/(2) = 0$$

The common denominator is 6.

$$20 - 2V_1 + V_2 - V_1 + 3V_3 - 3V_1 = 0$$

Equation 1:

$$-6V_1 + V_2 + 3V_3 = -20$$

$$I_4 + I_5 + I_6 = 0$$
$$(V_1 - V_2)/(R_2) - V_2/R_5 + (V_3 - V_2)/(R_4) = 0$$
$$(V_1 - V_2)/(6) - V_2/12 + (V_3 - V_2)/(4) = 0$$

The common denominator is 12.

$$2V_1 - 2V_2 - V_2 + 3V_3 - 3V_2 = 0$$

Equation 2:

$$2V_1 - 6V_2 + 3V_3 = 0$$

$$I_7 + I_8 + I_9 = 0$$
$$(V_1 - V_3)/(R_3) + (V_2 - V_3)/(R_4) + (E_2 - V_3)/(R_6) = 0$$
$$(V_1 - V_3)/(2) + (V_2 - V_3)/(4) + (15 - V_3)/(8) = 0$$

The common denominator is 8.

$$4V_1 - 4V_3 + 2V_2 - 2V_3 + 15 - V_3 = 0$$

Equation 3:

$$4V_1 + 2V_2 - 7V_3 = -15$$

Using determinants, solve for V_1.

$$V_1 = \frac{\begin{vmatrix} -20 & 1 & 3 \\ 0 & -6 & 3 \\ -15 & 2 & -7 \end{vmatrix}}{\begin{vmatrix} -6 & 1 & 3 \\ 2 & -6 & 3 \\ 4 & 2 & -7 \end{vmatrix}} = \frac{-840 - 45 + 0 - 270 + 120 + 0}{-252 + 12 + 12 + 72 + 36 + 14}$$

$$V_1 = -1{,}035/-106 = 9.76 \text{ volts}$$

$$I_1 = (E_1 - V_1)/R_1 = (10 - 9.76)/3 = 0.08 \text{ amp}$$

6

Superposition

In networks containing two or more sources, it is possible to avoid the use of simultaneous equations such as were used in the previous chapter. If you know how much currrent each source contributes to the individual branch current, you can find the branch current. The current that flows in any branch is equal to the algebraic sum of the currents flowing in the branch. The Superposition Theorem is useful in determining the current through or voltage drop across one branch. It is not to the student's advantage to use this theorem to solve for *all* currents and voltage drops in a network.

In order to determine the current flowing through or voltage drop across any component, proceed as follows:

1. Short out all sources except one. Find how much current this source contributes to the branch.

2. Repeat this process until all sources have been examined. For example, in Fig. 6.1, to determine the current flowing through R_4, first short out E_2 and find the amount of current flowing through R_4 due to E_1. Next short out E_1 and find the amount of current flowing through R_4 due to E_2.

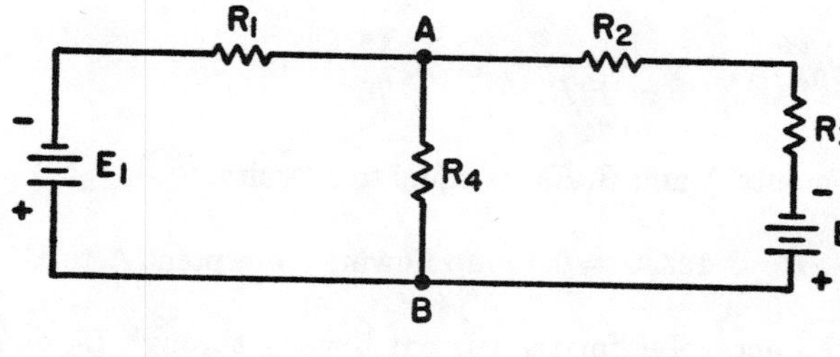

Fig. 6.1.

3. As soon as all the currents in the branch have been determined, take the algebraic sum of these currents to find the total current. If these currents flow in the same direction, add them. If they flow in the opposite direction, take the difference between the two.

4. Make sure that any resistance associated with a source is not removed from the circuit when the source is shorted out.

Problems

6.1 Find the current flowing through R_3 in Fig. 6.2.

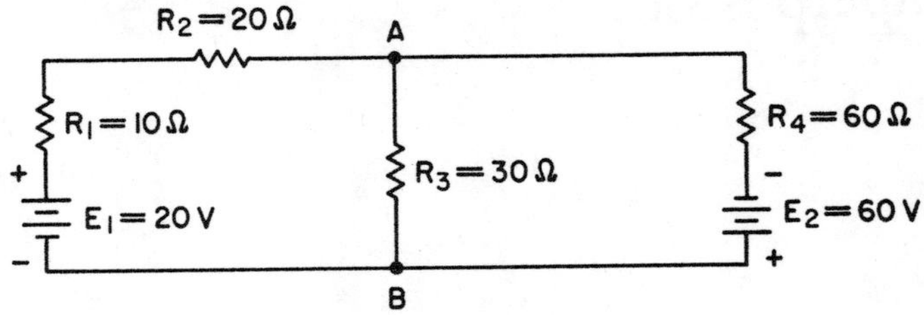

Fig. 6.2.

Answer

Short out source E_1 and solve for the current flowing through R_3 due to E_2 (see Fig. 6.2a). The series combination of R_1 and R_2 is in parallel with R_3 to form R_{AB}:

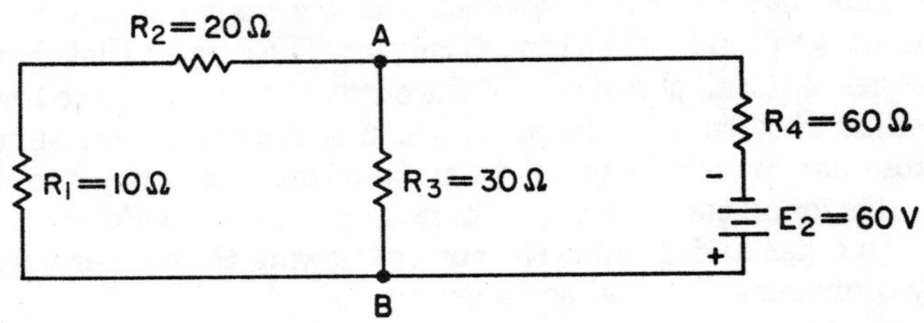

Fig. 6.2a.

$$R_{AB} = \frac{R_3(R_1 + R_2)}{R_3 + (R_1 + R_2)} = \frac{30(30)}{30 + 30} = 15 \text{ ohms}$$

You now have a series circuit consisting of R_{AB}, R_4, and E_2. E_{BA} can be found by using the law of proportionality.

$$E_{BA} = E_2 \frac{R_{AB}}{R_{AB} + R_4} = 60 \times \frac{15}{75} = 12 \text{ volts}$$

Since R_3 is across points A and B, E_{R_3} is equal to 12 volts.

$$I_{R3} = E_{R3}/R_3 = 12/30 = 0.4 \text{ amp flowing from point A to B}$$

Short out source E_2 and solve for the current flowing through R_3 due to E_1 (see Fig. 6.2b). R_4 is in parallel with R_3 to form R_{AB}:

$$R_{AB} = R_3R_4/(R_3 + R_4) = (30 \times 60)/(30 + 60) = 20 \text{ ohms}$$

You now have a series circuit consisting of R_1, R_2, R_{AB}, and E_1. E_{AB} can be found by using the law of proportionality.

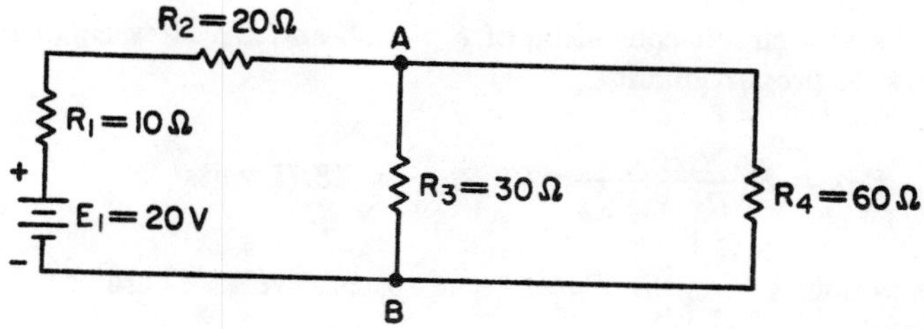

Fig. 6.2b.

$$E_{AB} = E_1 \frac{R_{AB}}{R_1 + R_2 + R_{AB}} = 20 \times \frac{20}{50} = 8 \text{ volts}$$

Since R_3 is across points A and B, E_{R3} is equal to 8 volts.

$$I_{R3} = E_{R3}/R_3 = 8/30 = 0.267 \text{ amp flowing from point B to A}$$

$$I_{R3} = 0.4 - 0.267 = 0.133 \text{ amp flowing from point A to B}$$

6.2 Find the current flowing through R_3 in Fig. 6.3.

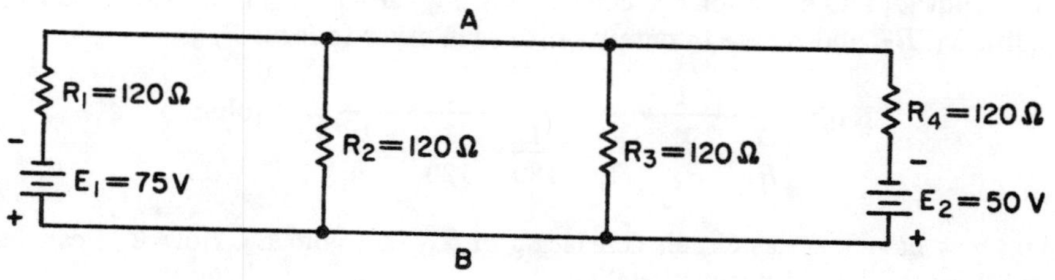

Fig. 6.3.

Answer

Short out E_2 and solve for the current flowing through R_3 due to E_1 (see Fig. 6.3a). R_2, R_3, and R_4 are in parallel with each other to form R_{AB}.

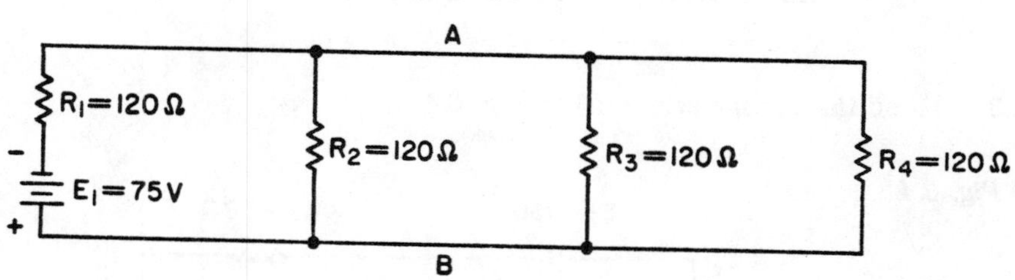

Fig. 6.3a.

$$R_{AB} = \frac{1}{\frac{1}{R_2} + \frac{1}{R_3} + \frac{1}{R_4}} = \frac{1}{\frac{1}{120} + \frac{1}{120} + \frac{1}{120}} = 40 \text{ ohms}$$

You now have a series circuit consisting of R_1, R_{AB}, and E_1. Now E_{BA} can be found by the law of proportionality.

$$E_{BA} = E_1 \frac{R_{AB}}{R_1 + R_{AB}} = 75 \times \frac{40}{160} = 18.75 \text{ volts}$$

Since R_3 is across points A and B, E_{R_3} is equal to 18.75 volts. Therefore,

$$I_{R3} = E_{R3}/R_3 = 18.75/120 = 0.156 \text{ amp flowing from A to B}$$

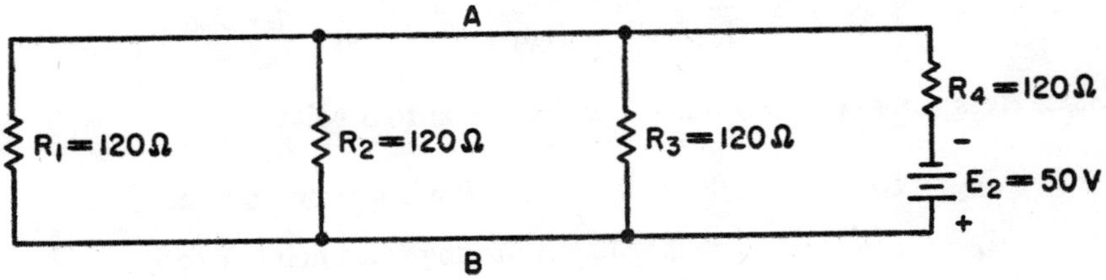

Fig. 6.3b.

Short out E_1 and solve for the current flowing through R_3 due to E_2 (see Fig. 6.3b). R_1, R_2, and R_3 are in parallel with each other to form R_{AB}.

$$R_{AB} = \frac{1}{\frac{1}{R_1} + \frac{1}{R_2} + \frac{1}{R_3}} = \frac{1}{\frac{1}{120} + \frac{1}{120} + \frac{1}{120}} = 40 \text{ ohms}$$

You now have a series circuit consisting of R_4, R_{AB}, and E_2. Now E_{BA} can be found by the law of proportionality.

$$E_{BA} = E_2 \frac{R_{AB}}{R_{AB} + R_4} = 50 \times \frac{40}{40 + 120} = 12.5 \text{ volts}$$

Since R_3 is across points A and B, E_{R_3} is equal to 12.5 volts.

$$I_{R3} = E_{R3}/R_3 = 12.5/120 = 0.104 \text{ amp flowing from A to B}$$

$$I_{R3} = 0.156 + 0.104 = 0.26 \text{ amp flowing from A to B}$$

6.3 Find the voltage across R_2 in Fig. 6.4.

Fig. 6.4.

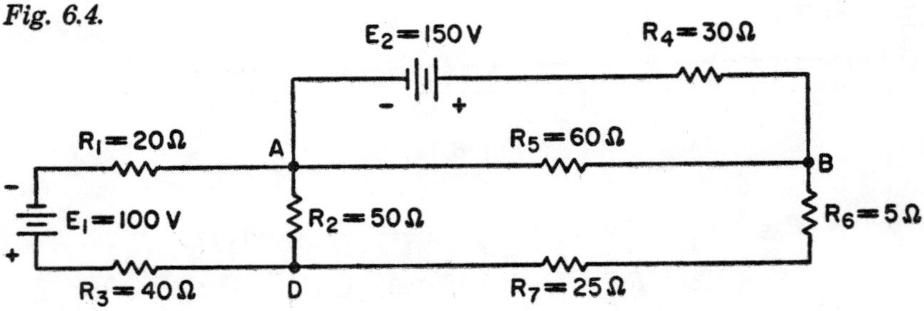

Answer

Short out source E_2 and find the voltage drop across R_2 due to E_1 (see Fig. 6.4a). R_4 and R_5 are in parallel with each other and form R_{AB}.

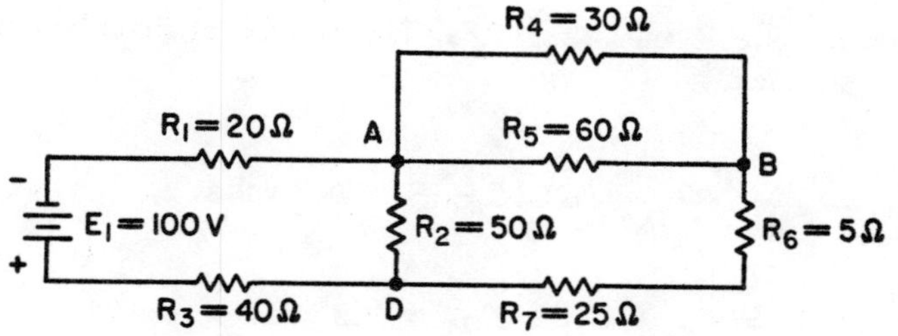

Fig. 6.4a.

$$R_{AB} = R_4R_5/(R_4 + R_5) = (30 \times 60)/(30 + 60) = 20 \text{ ohms}$$

The series combination of R_{AB}, R_6, and R_7 is in parallel with R_2 to form R_{AD}.

$$R_{AD} = \frac{(R_{AB} + R_6 + R_7)R_2}{(R_{AB} + R_6 + R_7) + R_2} = \frac{50 \times 50}{50 + 50} = 25 \text{ ohms}$$

You now have a series circuit consisting of R_1, R_{AD}, R_3, and E_1. E_{DA} can now be found by the law of proportionality.

$$E_{DA} = E_1 \frac{R_{AD}}{R_{AD} + R_1 + R_3} = 100 \times \frac{25}{85} = 29.4 \text{ volts}$$

Since R_2 is across points A and D, E_{R_2} due to E_1 is equal to 29.4 volts.

Now short out source E_1 and find the voltage drop across R_2 due to E_2 (see Fig. 6.4b). The series combination of R_1 and R_3 is in parallel with R_2 to form R_{AD}.

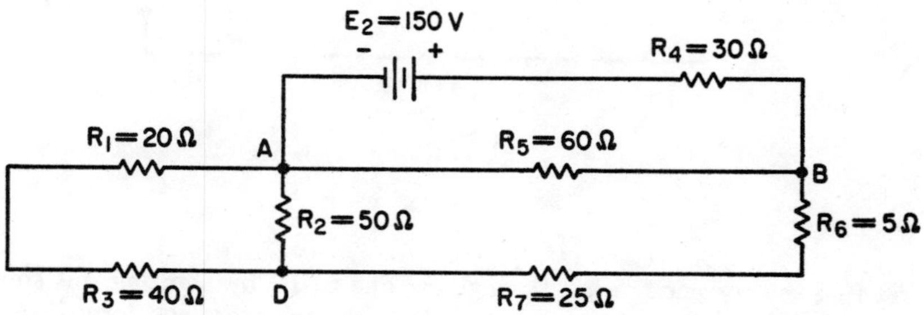

Fig. 6.4b.

$$R_{AD} = \frac{(R_1 + R_3)R_2}{(R_1 + R_3) + R_2} = \frac{60 \times 50}{60 + 50} = 27.3 \text{ ohms}$$

The series combination of R_{AD}, R_7, and R_6 is in parallel with R_5 to form R_{AB}.

$$R_{AB} = \frac{(R_{AD} + R_7 + R_6)R_5}{(R_{AD} + R_7 + R_6) + R_5} = \frac{57.3 \times 60}{57.3 + 60} = 29.3 \text{ ohms}$$

You now have a series circuit consisting of R_4, R_{AB}, and E_2. E_{BA} can be found by the law of proportionality:

$$E_{BA} = E_2 \frac{R_{AB}}{R_{AB} + R_4} = 150 \times \frac{29.2}{59.2} = 74.2 \text{ volts}$$

$$E_{BA} = E_{R5} = E_{R6} + E_{R7} + E_{DA} = 74.2 \text{ volts}$$

Now find E_{DA} using the law of proportionality.

$$E_{DA} = E_{BA} \frac{R_{AD}}{R_{AD} + R_7 + R_6} = 74 \times \frac{27.3}{57.3} = 35.3 \text{ volts}$$

Since R_2 is across points A and D, E_{R2} due to E_2 is equal to 35.3 volts. The total voltage across R_2 is thus

$$E_{R2} = 35.3 + 29.4 = 64.7 \text{ volts}$$

6.4 Find the current flowing through R_2 in Fig. 6.5.

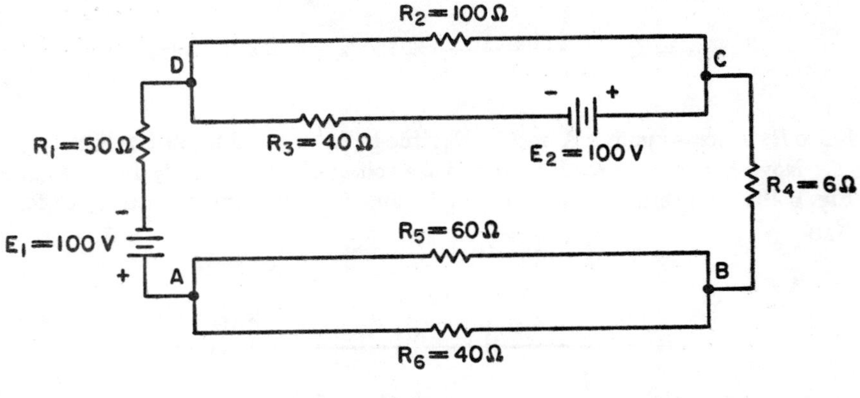

Fig. 6.5.

Answer

First find the current flowing through R_2 due to E_1 by shorting out source E_2 (see Fig. 6.5a). R_2 and R_3 are in parallel with each other and form R_{DC}:

$$R_{DC} = R_2R_3/(R_2 + R_3) = (100 \times 40)/(100 + 40) = 28.6 \text{ ohms}$$

R_5 and R_6 are in parallel with each other and form R_{AB}.

$$R_{AB} = R_5R_6/(R_5 + R_6) = (60 \times 40)/(60 + 40) = 24 \text{ ohms}$$

You now have a series circuit consisting of R_1, R_{DC}, R_4, R_{AB}, and E_1. E_{CD} can now be found by using the law of proportionality.

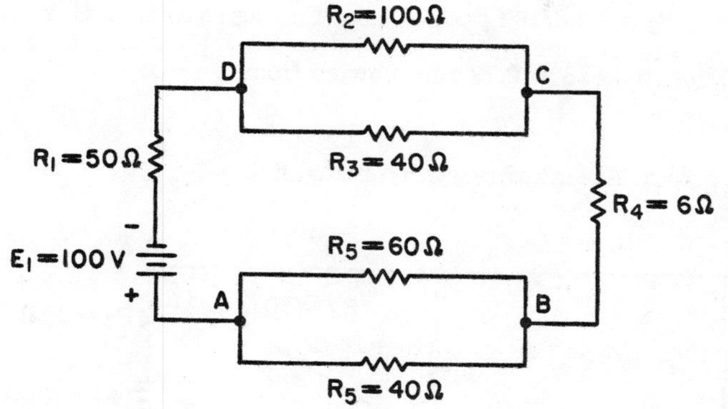

Fig. 6.5a.

$$E_{CD} = E_1 \frac{R_{DC}}{R_{DC} + R_1 + R_4 + R_{AB}} = 100 \times \frac{28.6}{108.6} = 26.3 \text{ volts}$$

Since R_2 is across points C and D, E_{R_2} is equal to 26.4 volts.

$$I_{R2} = E_{R2}/R_2 = 26.4/100 = 0.264 \text{ amp flowing from D to C}$$

Now find the current flowing through R_2 due to E_2 by shorting out source E_1 (see Fig. 6.5b). R_5 and R_6 are in parallel, and as calculated previously in this problem, R_{AB} equals 24 ohms. The series combination of R_1, R_{AB}, and R_4 are in parallel with R_2 to form R_{DC}:

$$R_{DC} = \frac{(R_1 + R_{AB} + R_4)R_2}{(R_1 + R_{AB} + R_4) + R_2} = \frac{80 \times 100}{80 + 100} = 44.4 \text{ ohms}$$

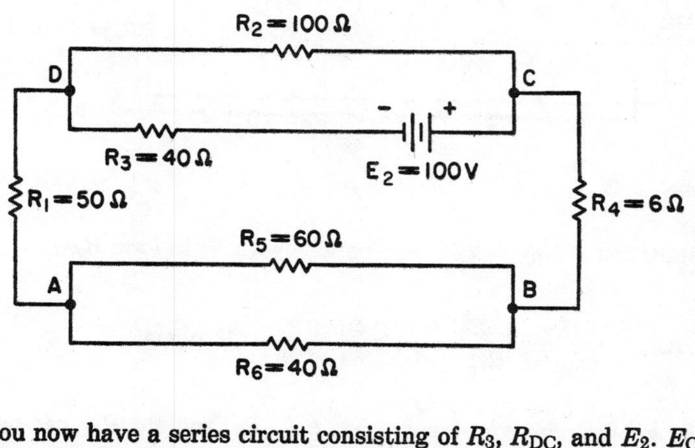

Fig. 6.5b.

You now have a series circuit consisting of R_3, R_{DC}, and E_2. E_{CD} can now be found by using the law of proportionality.

$$E_{CD} = E_2 \frac{R_{DC}}{R_{DC} + R_3} = 100 \times \frac{44.4}{84.4} = 52.6 \text{ volts}$$

Since R_2 is across points D and C, E_{R_2} is equal to 52.6 volts.

$$I_{R2} = E_{R2}/R_2 = 52.6/100 = 0.526 \text{ amp flowing from D to C}$$

$$I_{R2} = 0.264 + 0.526 = 0.79 \text{ amp flowing from D to C}$$

6.5 Find the current flowing through R_3 in Fig. 6.6.

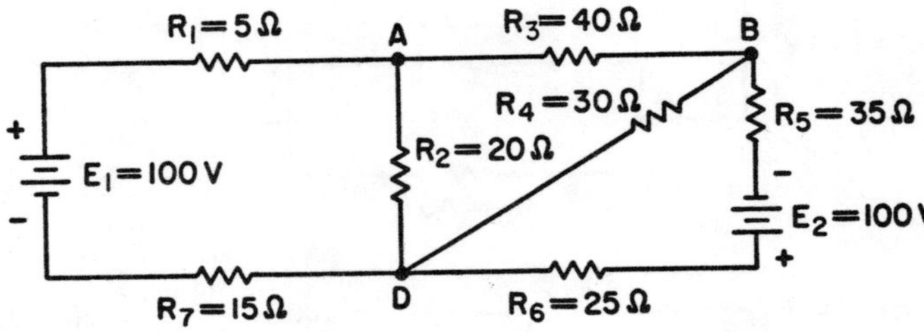

Fig. 6.6.

Answer

First find the current flowing through R_3 due to E_1 by shorting out source E_2 (see Fig. 6.6a). The series combination of R_5 and R_6 is in parallel with R_4 to form R_{BD}:

$$R_{BD} = \frac{(R_5 + R_6)R_4}{(R_5 + R_6) + R_4} = \frac{60 \times 30}{60 + 30} = 20 \text{ ohms}$$

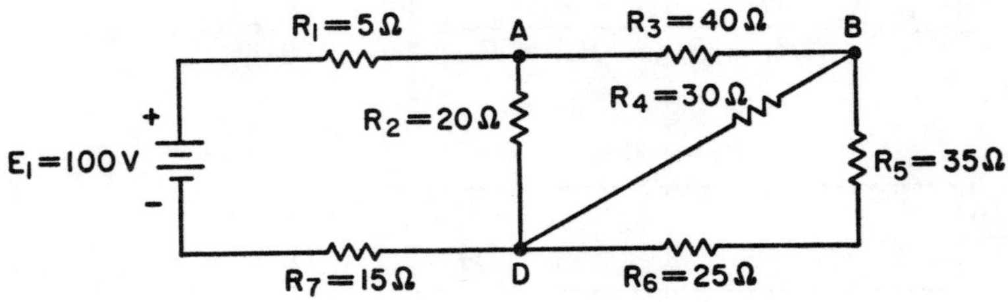

Fig. 6.6a.

The series combination of R_{BD} and R_3 is in parallel with R_2 to form R_{AD}.

$$R_{AD} = \frac{(R_{BD} + R_3)R_2}{(R_{BD} + R_3) + R_2} = \frac{60 \times 20}{60 + 20} = 15 \text{ ohms}$$

You now have a series circuit consisting of R_1, R_{AD}, R_7, and E_1. E_{AD} can now be found by using the law of proportionality.

$$E_{AD} = E_1 \frac{R_{AD}}{R_1 + R_{AD} + R_7} = 100 \times \frac{15}{35} = 42.8 \text{ volts}$$

$$E_{AD} = E_{R2} = E_{R3} + E_{BD} = 42.8 \text{ volts}$$

Using Ohm's law, the current flowing through R_3 due to E_1 is

$$I_{R3} = (E_{R3} + E_{BD})/(R_3 + R_{BD}) = 42.8/60 = 0.713 \text{ amp flowing from B to A}$$

Now find the current flowing through R_3 due to E_2 by shorting out source E_1 (see Fig. 6.6b). The series combination of R_1 and R_7 is in parallel with R_2 to form R_{AD}:

$$R_{AD} = \frac{(R_1 + R_7)R_2}{(R_1 + R_7) + R_2} = \frac{20 \times 20}{20 + 20} = 10 \text{ ohms}$$

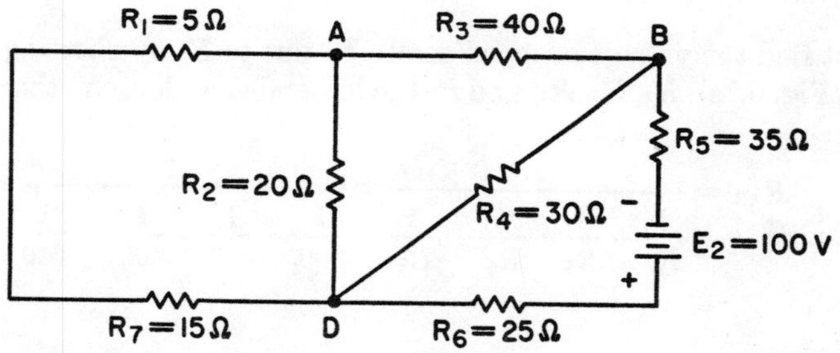

Fig. 6.6b.

The series combination of R_{AD} and R_3 is in parallel with R_4 to form R_{BD}.

$$R_{BD} = \frac{(R_{AD} + R_3)R_4}{(R_{AD} + R_3) + R_4} = \frac{50 \times 30}{50 + 30} = 18.75 \text{ ohms}$$

You now have a series circuit consisting of R_5, R_6, R_{BD}, and E_2. E_{DB} can now be found by using the law of proportionality.

$$E_{DB} = E_2 \frac{R_{BD}}{R_5 + R_6 + R_{BD}} = 100 \times \frac{18.75}{78.75} = 23.8 \text{ volts}$$

$$E_{DB} = E_{R4} = E_{R3} + E_{DA} = 23.8 \text{ volts}$$

The current flowing through R_3 due to E_2 is thus

$$I_{R3} = (E_{R3} + E_{DA})/(R_3 + R_{AD}) = 23.8/50 = 0.476 \text{ amp flowing from B to A}$$

The total current flowing through R_3 is thus

$$I_{R3} = 0.713 + 0.476 = 1.189 \text{ amps flowing from B to A}$$

6.6 Find the voltage across R_3 in Fig. 6.7.

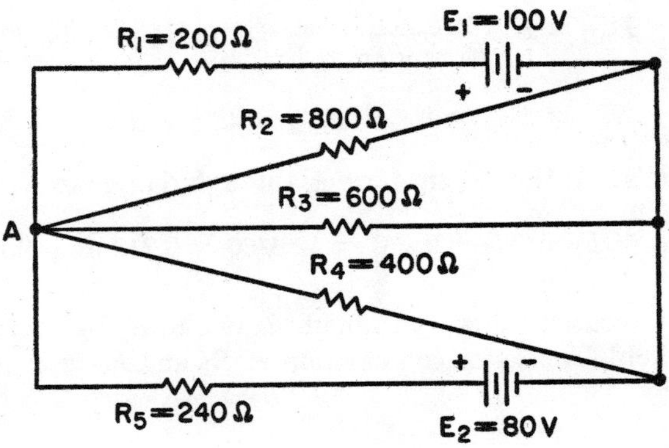

Fig. 6.7.

Answer

First find the voltage dropped across R_3 due to E_1 by shorting out source E_2 (see Fig. 6.7a). R_2, R_3, R_4, and R_5 are in parallel with each other to form R_{AB}.

$$R_{AB} = \frac{1}{\frac{1}{R_2}+\frac{1}{R_3}+\frac{1}{R_4}+\frac{1}{R_5}} = \frac{1}{\frac{1}{800}+\frac{1}{600}+\frac{1}{400}+\frac{1}{240}} = 104 \text{ ohms}$$

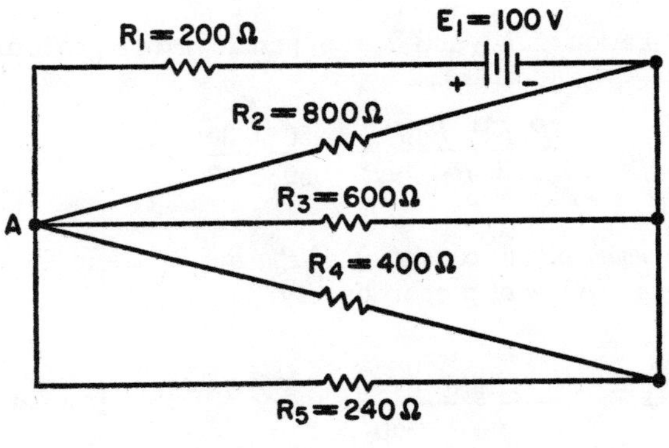

Fig. 6.7a.

You now have a series circuit consisting of R_{AB}, R_1, and E_1. E_{AB} can be found by using the law of proportionality.

$$E_{AB} = E_1 \frac{R_{AB}}{R_{AB} + R_1} = 100 \times \frac{104}{304} = 34.2 \text{ volts}$$

$$E_{AB} = E_{R2} = E_{R3} = E_{R4} = E_{R5} = 34.2 \text{ volts}$$

Thus the voltage dropped across R_3 due to E_1 is 34.2 volts. Now find the voltage dropped across R_3 due to E_2 by shorting out source E_1 (see Fig. 6.7b). R_1, R_2, R_3, and R_4 are in parallel with each other to form R_{AB}.

$$R_{AB} = \frac{1}{\frac{1}{R_1}+\frac{1}{R_2}+\frac{1}{R_3}+\frac{1}{R_4}} = \frac{1}{\frac{1}{200}+\frac{1}{800}+\frac{1}{600}+\frac{1}{400}} = 96 \text{ ohms}$$

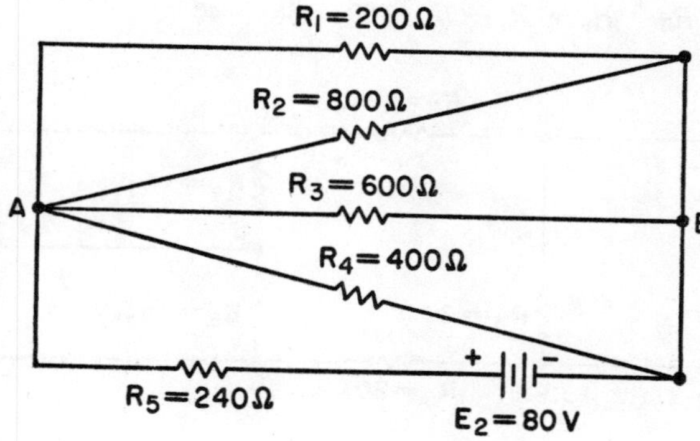

Fig. 6.7b.

You now have a series circuit consisting of R_{AB}, R_5, and E_2. E_{AB} can be found by using the law of proportionality.

$$E_{AB} = E_2 \frac{R_{AB}}{R_{AB} + R_5} = 80 \times \frac{96}{336} = 22.9 \text{ volts}$$

$$E_{AB} = E_{R1} = E_{R2} = E_{R3} = E_{R4} = 22.9 \text{ volts}$$

Thus the voltage dropped across R_3 due to E_2 is 22.9 volts, and the total voltage dropped across it is

$$E_{R3} = 34.2 + 22.9 = 57.1 \text{ volts}$$

6.7 Find the current flowing through R_3 in Fig. 6.8.

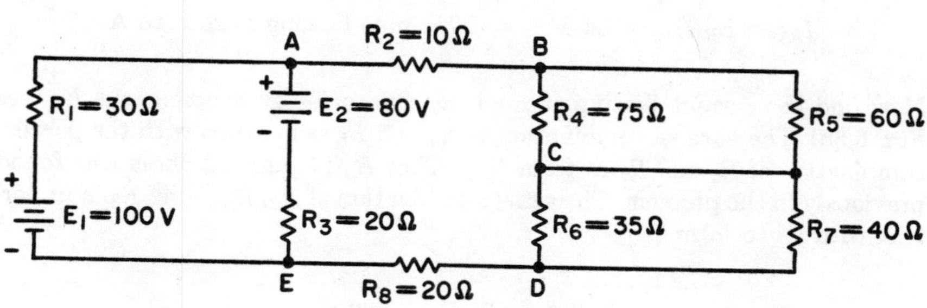

Fig. 6.8.

Answer

First find the current flowing through R_3 due to E_1 by shorting out source E_2 (see Fig. 6.8a). The parallel combination of R_4 and R_5 is in series with the parallel combination of R_6 and R_7 to form R_{BD}.

$$R_{BD} = \frac{R_4R_5}{R_4 + R_5} + \frac{R_6R_7}{R_6 + R_7} = \frac{75 \times 60}{75 + 60} + \frac{35 \times 40}{35 + 40} = 33.3 + 18.7 = 52 \text{ ohms}$$

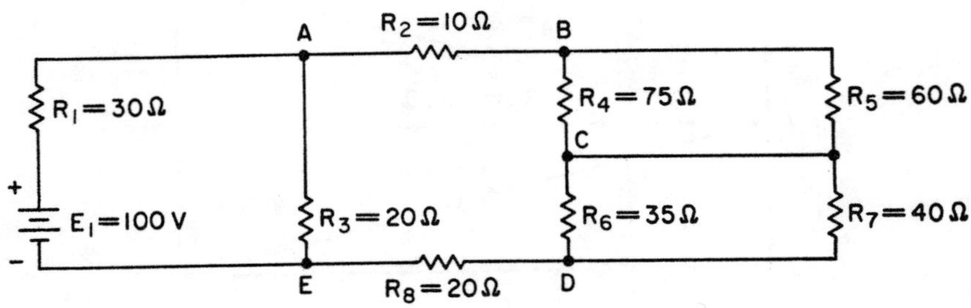

Fig. 6.8a.

The series combination of R_2, R_{BD}, and R_8 is in parallel with R_3 to form R_{AE}.

$$R_{AE} = \frac{R_3(R_2 + R_{BD} + R_8)}{R_3 + (R_2 + R_{BD} + R_8)} = \frac{20 \times 82}{20 + 82} = 16.1 \text{ ohms}$$

You now have a series circuit consisting of R_{AE}, R_1, and E_1. E_{AE} can now be found by using the law of proportionality:

$$E_{AE} = E_1 \frac{R_{AE}}{R_{AE} + R_1} = 100 \times \frac{16.1}{46.1} = 34.9 \text{ volts}$$

$$E_{AE} = E_{R3} = E_{R2} + E_{RBD} + E_{R8} = 34.9 \text{ volts}$$

The voltage drop across R_3 is thus 34.9 volts. The current flowing through R_3 due to E_1 is therefore

$$I_{R3} = E_{R3}/R_3 = 34.9/20 = 1.75 \text{ amps flowing from E to A}$$

Now find the current flowing through R_3 due to E_2 by shorting out E_1 (see Fig. 6.8b). The parallel combination of R_4 and R_5 is in series with the parallel combination of R_6 and R_7 to form R_{BD}. That R_{BD} equals 52 ohms was found previously in the problem. The series combination of R_2, R_{BD}, and R_8 is in parallel with R_1 to form R_{AE}.

$$R_{AE} = \frac{(R_2 + R_{BD} + R_8)R_1}{(R_2 + R_{BD} + R_8) + R_1} = \frac{82 \times 30}{82 + 30} = 22 \text{ ohms}$$

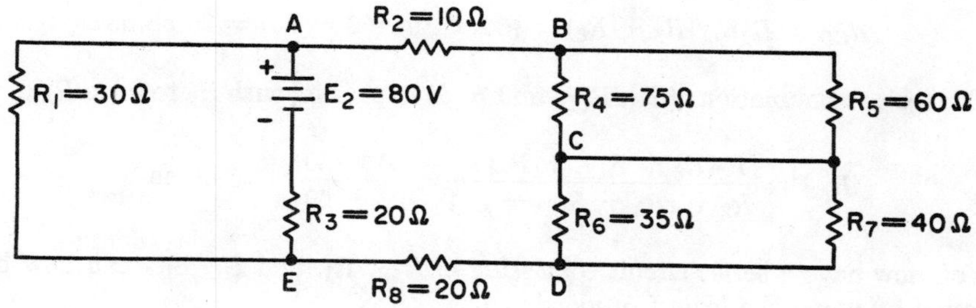

Fig. 6.8b.

You now have a series circuit consisting of R_{AE}, R_3, and E_2. The current flowing through R_3 due to E_2 is therefore

$$I_{R3} = E_2/(R_{AE} + R_3) = 80/42 = 1.90 \text{ amps flowing from A to E}$$

The total current flowing through I_{R3} is then

$$I_{R3} = 1.9 - 1.75 = 0.15 \text{ amp flowing from A to E}$$

6.8 Find the voltage across R_2 in Fig. 6.9.

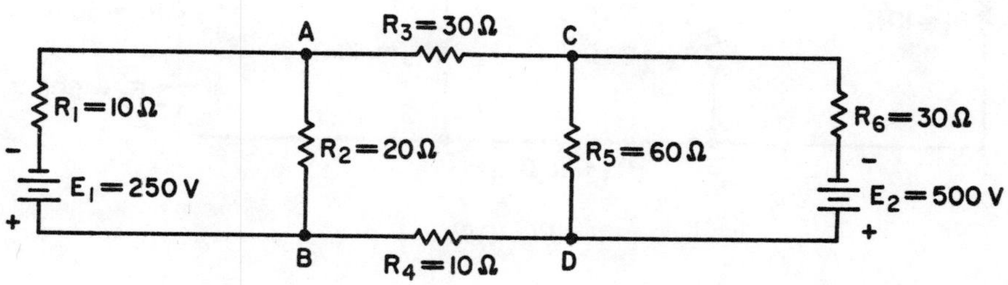

Fig. 6.9.

Answer

First find the voltage across R_2 due to E_1 by shorting out E_2 (see Fig. 6.9a). R_5 and R_6 are in parallel with each other to form R_{CD}:

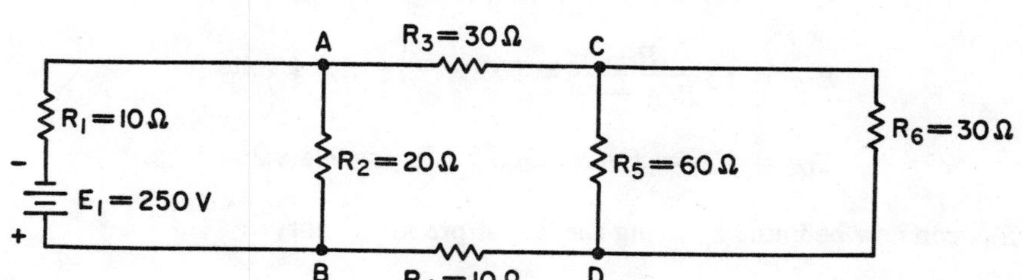

Fig. 6.9a.

$$R_{CD} = R_5R_6/(R_5 + R_6) = (60 \times 30)/(60 + 30) = 20 \text{ ohms}$$

The series combination of R_3, R_{CD}, and R_4 is in parallel with R_2 to form R_{AB}.

$$R_{AB} = \frac{R_2(R_3 + R_{CD} + R_4)}{R_2 + (R_3 + R_{CD} + R_4)} = \frac{20 \times 60}{20 + 60} = 15 \text{ ohms}$$

You now have a series circuit consisting of R_{AB}, R_1, and E_1. E_{BA} can now be found by using the law of proportionality.

$$E_{BA} = E_1\frac{R_{AB}}{R_1 + R_{AB}} = 250 \times \frac{15}{25} = 150 \text{ volts}$$

Since R_2 is across points A and B, E_{R2} due to E_1 is equal to 150 volts.

Now find the voltage across R_2 due to E_2 by shorting out source E_1 (see Fig. 6.9b). R_1 and R_2 are in parallel with each other to form R_{AB}.

$$R_{AB} = R_1R_2/(R_1 + R_2) = (10 \times 20)/(10 + 20) = 6.67 \text{ ohms}$$

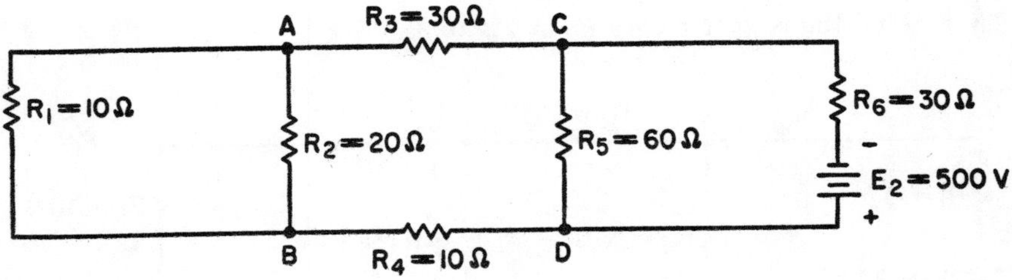

Fig. 6.9b.

The series combination of R_{AB}, R_3, and R_4 is in parallel with R_5 to form R_{CD}.

$$R_{CD} = \frac{(R_{AB} + R_3 + R_4)R_5}{(R_{AB} + R_3 + R_4) + R_5} = \frac{46.67 \times 60}{46.67 + 60} = 26.3 \text{ ohms}$$

You now have a series circuit consisting of R_{CD}, R_6, and E_2. E_{DC} can now be found by using the law of proportionality.

$$E_{DC} = E_2\frac{R_{CD}}{R_{CD} + R_6} = 500 \times \frac{26.3}{56.3} = 234 \text{ volts}$$

$$E_{DC} = E_{R5} = E_{R3} + E_{BA} + E_{R_4} = 234 \text{ volts}$$

E_{BA} can now be found by using the law of proportionality.

$$E_{BA} = E_{DC}\frac{R_{AB}}{R_3 + R_{AB} + R_4} = 234 \times \frac{6.67}{46.67} = 33.5 \text{ volts}$$

Since R_2 is across points A and B, E_{R2} due to E_2 is equal to 33.5 volts. The total voltage across R_2 is thus

$$E_{R2} = 150 + 33.5 = 183.5 \text{ volts}$$

6.9 Find the voltage across R_1 in Fig. 6.10.

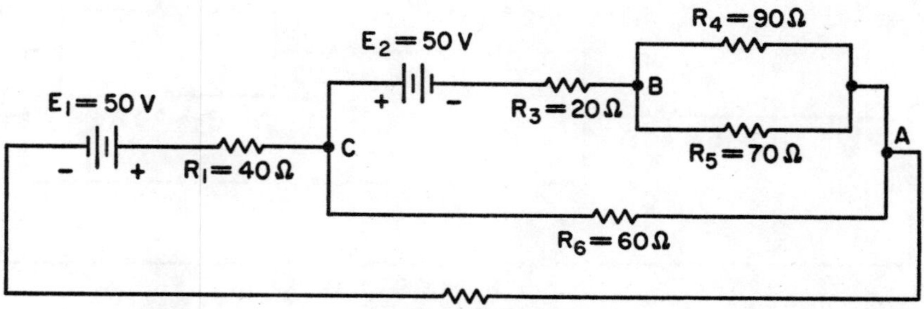

Fig. 6.10.

Answer

First find the voltage dropped across R_1 due to E_1 by shorting out source E_2 (Fig. 6.10a). R_4 and R_5 are in parallel with each other to form R_{AB}:

$$R_{AB} = R_4R_5/(R_4 + R_5) = (90 \times 70)/(90 + 70) = 39.4 \text{ ohms}$$

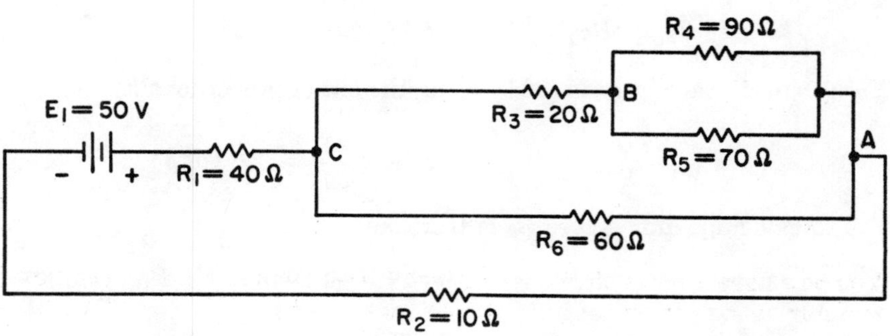

Fig. 6.10a.

The series combination of R_{AB} and R_3 is in parallel with R_6 to form R_{AC}.

$$R_{AC} = \frac{(R_{AB} + R_3)R_6}{(R_{AB} + R_3) + R_6} = \frac{59.4 \times 60}{59.4 + 60} = 29.8 \text{ ohms}$$

You now have a series circuit consisting of R_1, R_{AC}, R_2, and E_1. The voltage drop across R_1 due to E_1 can now be found by using the law of proportionality.

$$E_{R1} = E_1 \frac{R_1}{R_1 + R_{AC} + R_2} = 50 \times \frac{40}{79.8} = 25.1 \text{ volts}$$

Find the voltage drop across R_1 due to E_2 by shorting out source E_1 (Fig. 6.10b). The series combination of R_1 and R_2 is in parallel with R_6 to form R_{AC}.

Fig. 6.10b.

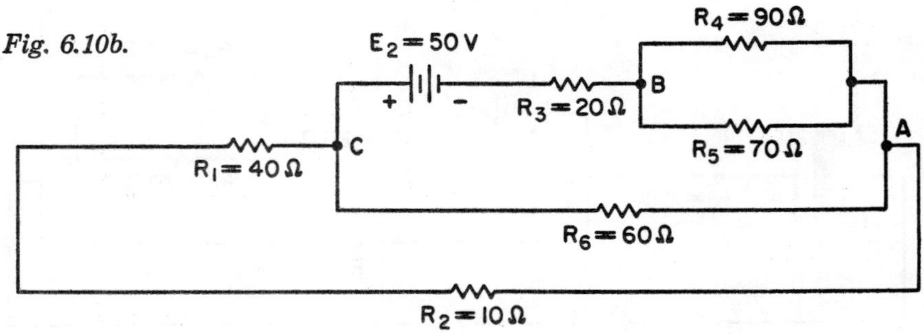

$$R_{AC} = \frac{(R_1 + R_2)R_6}{(R_1 + R_2) + R_6} = \frac{50 \times 60}{50 + 60} = 27.3 \text{ ohms}$$

R_4 and R_5 are in parallel with each other to form R_{AB}. It was found previously in the problem that $R_{AB} = 39.4$ ohms. You now have a series circuit consisting of R_{AB}, R_{AC}, R_3, and E_2. E_{CA} can now be found by using the law of proportionality.

$$E_{CA} = E_2 \frac{R_{AC}}{R_{AB} + R_{AC} + R_3} = 50 \times \frac{27.3}{86.7} = 15.78 \text{ volts}$$

$$E_{CA} = E_{R6} = E_{R2} + E_{R1} = 15.78 \text{ volts}$$

E_{R_1} due to E_2 can now be found by using the law of proportionality.

$$E_{R1} = E_{CA} \frac{R_1}{R_1 + R_2} = 15.78 \times \frac{40}{50} = 12.6 \text{ volts}$$

The total voltage drop across E_2 is therefore

$$E_{R1} = 25.1 - 12.6 = 12.5 \text{ volts}$$

6.10 Find the voltage across R_1 in Fig. 6.11.

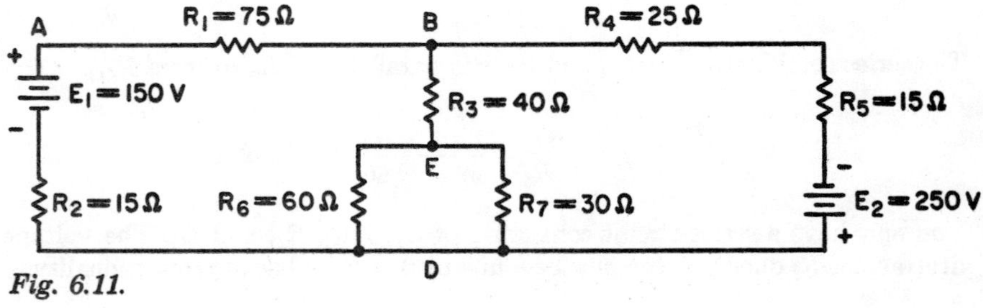

Fig. 6.11.

Answer

First find the voltage dropped across R_1 due to E_1 by shorting out source E_2 (see Fig. 6.11a). R_6 and R_7 are in parallel with each other to form R_{ED}.

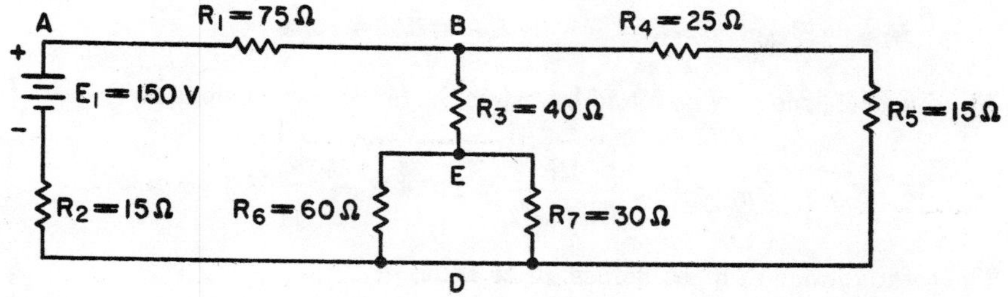

Fig. 6.11a.

$$R_{ED} = R_6R_7/(R_6 + R_7) = (60 \times 30)/(60 + 30) = 20 \text{ ohms}$$

The series combination of R_4 and R_5 is in parallel with the series combination of R_3 and R_{ED} to form R_{BD}.

$$R_{BD} = \frac{(R_3 + R_{ED})(R_4 + R_5)}{(R_3 + R_{ED}) + (R_4 + R_5)} = \frac{60 \times 40}{60 + 40} = 24 \text{ ohms}$$

You now have a series circuit consisting of R_1, R_{BD}, R_2, and E_1. E_{R1} due to E_1 can now be found by using the law of proportionality.

$$E_{AB} = E_{R1} = E_1 \frac{R_1}{R_1 + R_{BD} + R_2} = 150 \times \frac{75}{114} = 98.7 \text{ volts}$$

Now find the voltage dropped across R_1 due to E_2 by shorting out source E_1 (see Fig. 6.11b). R_6 and R_7 are in parallel with each other to form R_{ED}, which was found to be 20 ohms previously in the problem. Moreover, the series combination of R_1 and R_2 is in parallel with the series combination of R_3 and R_{ED} to form R_{BD}.

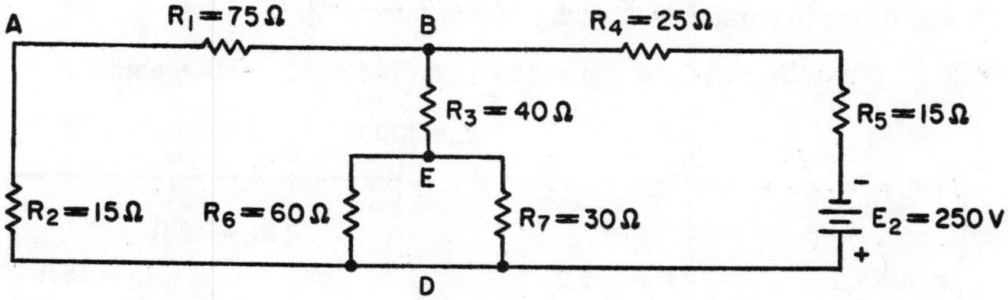

Fig. 6.11b.

$$R_{BD} = \frac{(R_1 + R_2)(R_3 + R_{ED})}{(R_1 + R_2) + (R_3 + R_{ED})} = \frac{(90)(60)}{(90) + (60)} = 36 \text{ ohms}$$

You now have a series circuit consisting of R_{BD}, R_4, R_5, and E_2, and you can find E_{DB} by using the law of proportionality.

$$E_{DB} = E_2 \frac{R_{BD}}{R_{BD} + R_4 + R_5} = 250 \times \frac{36}{76} = 118.4 \text{ volts}$$

$$E_{DB} = E_{R3} + E_{RED} = E_{R1} + E_{R2} = 118.4 \text{ volts}$$

E_{R_1} due to E_2 can now be found by using the law of proportionality.

$$E_{AB} = E_{R1} = E_{DB} \frac{R_1}{R_1 + R_2} = 118.4 \times \frac{75}{90} = 98.7 \text{ volts}$$

The total voltage dropped across R_1 is thus

$$E_{R1} = 98.7 + 98.7 = 197.4 \text{ volts}$$

6.11 Find the current flowing through R_5 in Fig. 6.12.

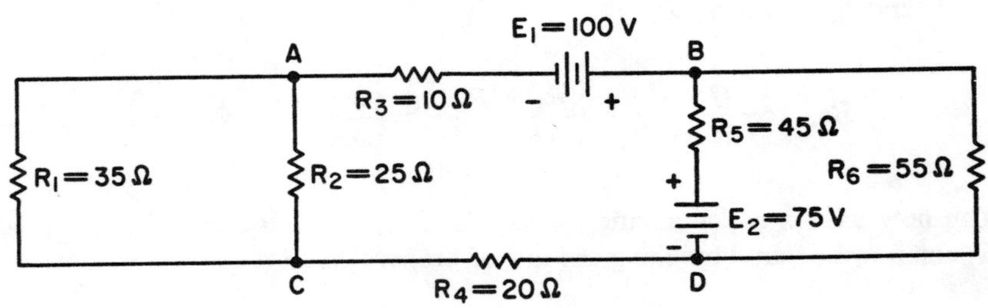

Fig. 6.12.

Answer

First find the current flowing through R_5 due to E_1 by shorting out source E_2 (see Fig. 6.12a). R_5 and R_6 are in parallel with each other to form R_{BD}.

$$R_{BD} = R_5R_6/(R_5 + R_6) = (45 \times 55)/(45 + 55) = 24.8 \text{ ohms}$$

R_1 and R_2 are in parallel with each other to form R_{AC}:

$$R_{AC} = R_1R_2/(R_1 + R_2) = (35 \times 25)/(35 + 25) = 14.6 \text{ ohms}$$

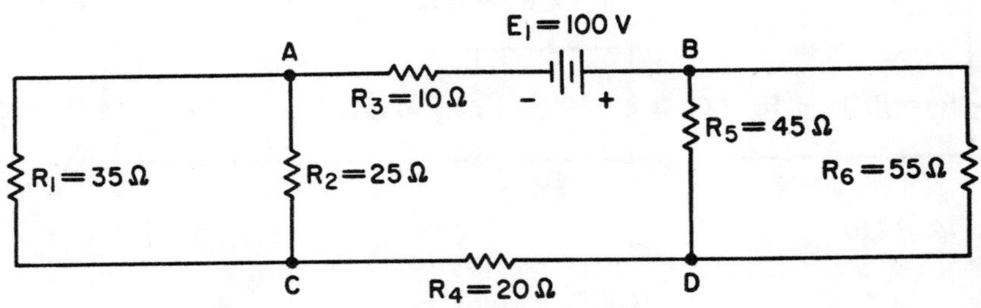

Fig. 6.12a.

You now have a series circuit consisting of R_{AC}, R_3, R_4, R_{BD}, and E_1. E_{BD} can now be found by using the law of proportionality.

$$E_{BD} = E_1 \frac{R_{BD}}{R_{AC} + R_3 + R_4 + R_{BD}} = 100 \times \frac{24.8}{69.4} = 35.8 \text{ volts}$$

$$E_{BD} = E_{R5} = E_{R6} = E_{R1} = E_{R2} = 35.8 \text{ volts}$$

The current flowing through R_5 due to E_1 is thus

$$I_{R5} = E_{R5}/R_5 = 35.8/45 = 0.796 \text{ amp flowing from D to B}$$

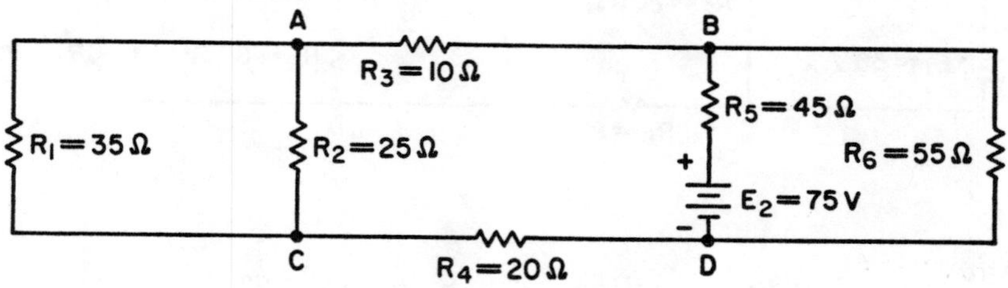

Fig. 6.12b.

Now find the current flowing through R_5 due to E_2 by shorting out source E_1 (see Fig. 6.12b). R_1 and R_2 are in parallel with each other to form R_{AC}, which was found to be 14.6 ohms previously in the problem. Moreover, the series combination of R_{AC}, R_3, and R_4 is in parallel with R_6 to form R_{BD}.

$$R_{BD} = \frac{(R_6)(R_{AC} + R_3 + R_4)}{(R_6) + (R_{AC} + R_3 + R_4)} = \frac{(55)(44.6)}{(55) + (44.6)} = 24.5 \text{ ohms}$$

You now have a series circuit consisting of R_{BD}, R_5, and E_2. The current flowing through R_5 due to E_2 is therefore

$$I_{R5} = E_2/(R_{BD} + R_5) = 75/69.5 = 1.08 \text{ amps flowing from B to D}$$

The total current flowing through R_5 is thus

$$I_{R5} = 1.08 - 0.796 = 0.284 \text{ amp flowing from B to D}$$

6.12 Find the voltage across R_8 in Fig. 6.13.

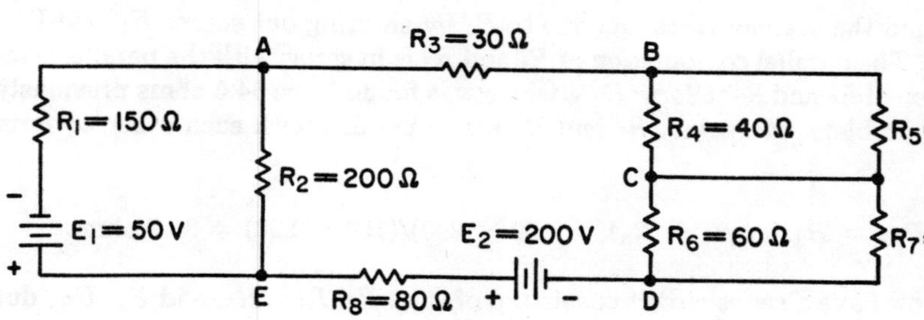

Fig. 6.13.

Answer

First find the voltage across R_8 due to E_1 by shorting out source E_2 (see Fig. 6.13a). The parallel combination of R_4 and R_5 is in series with the parallel combination of R_6 and R_7 to form R_{BD}.

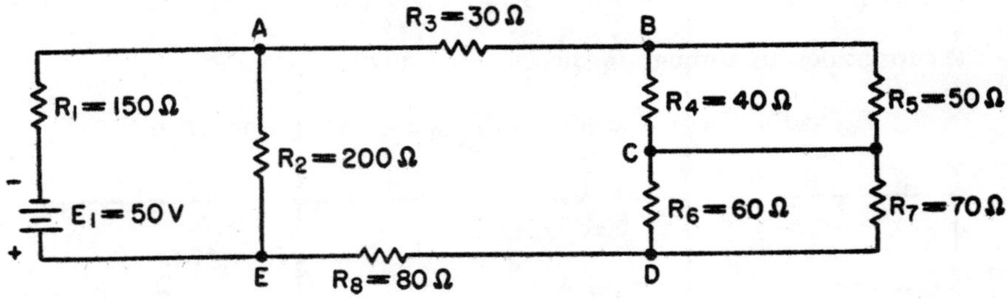

Fig. 6.13a.

$$R_{BD} = \frac{R_4R_5}{R_4 + R_5} + \frac{R_6R_7}{R_6 + R_7} = \frac{40 \times 50}{40 + 50} + \frac{60 \times 70}{60 + 70} = 22.2 + 32.3$$
$$= 54.5 \text{ ohms}$$

The series combination of R_3, R_{BD}, and R_8 is in parallel with R_2 to form R_{AE}.

$$R_{AE} = \frac{(R_3 + R_{BD} + R_8)R_2}{(R_3 + R_{BD} + R_8) + R_2} = \frac{164.5 \times 200}{164.5 + 200} = 90.3 \text{ ohms}$$

You now have a series circuit consisting of R_{AE}, R_1, and E_1. E_{EA} can now be found by using the law of proportionality.

$$E_{EA} = E_1 \frac{R_{AE}}{R_{AE} + R_1} = 50 \times \frac{90.3}{240.3} = 18.8 \text{ volts}$$

$$E_{EA} = E_{R2} = E_{R3} + E_{DB} + E_{R8} = 18.8 \text{ volts}$$

E_{R8} due to E_1 can now be found by using the law of proportionality.

$$E_{R8} = E_{EA} \frac{R_8}{R_8 + R_3 + R_{BD}} = 18.8 \times \frac{80}{164.5} = 9.14 \text{ volts}$$

Now find the voltage across R_8 due to E_2 by shorting out source E_1 (see Fig. 6.13b). The parallel combination of R_4 and R_5 is in series with the parallel combination of R_6 and R_7 to form R_{BD}, which was found to be 54.5 ohms previously in the problem. Moreover, R_1 and R_2 are in parallel with each other to form R_{AE}.

$$R_{AE} = R_1R_2/(R_1 + R_2) = (150 \times 200)/(150 + 200) = 85.7 \text{ ohms}$$

You now have a series circuit consisting of R_{AE}, R_3, R_{BD}, R_8, and E_2. E_{R8} due to E_2 can now be found by using the law of proportionality.

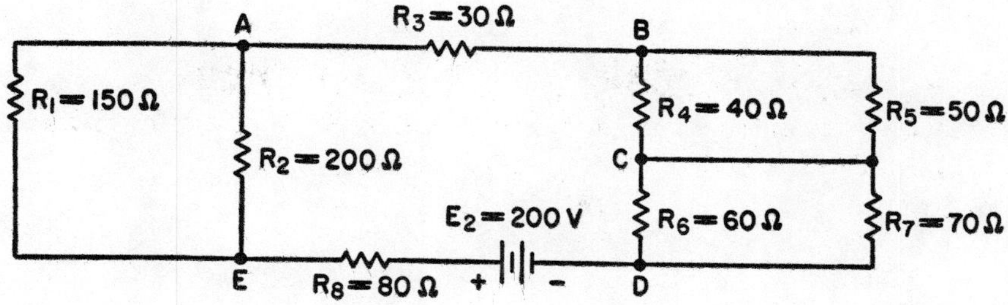

Fig. 6.13b.

$$E_{R8} = E_2 \frac{R_8}{R_8 + R_{AE} + R_{BD} + R_3} = 200 \times \frac{80}{250.2} = 64 \text{ volts}$$

The total voltage dropped across R_8 is thus

$$E_{R8} = 64 - 9.14 = 54.86 \text{ volts}$$

7

Millman's Theorem

Millman's theorem makes it possible to obtain the desired solution by writing simply one equation. It will not be necessary to write any simultaneous equations or to replace any part of the circuit.

Millman's theorem states that in any circuit containing two or more parallel voltage sources (and their internal resistance) the resultant voltage is equal to the ratio of the algebraic sum of the short circuit currents divided by the algebraic sum of the internal admittances.

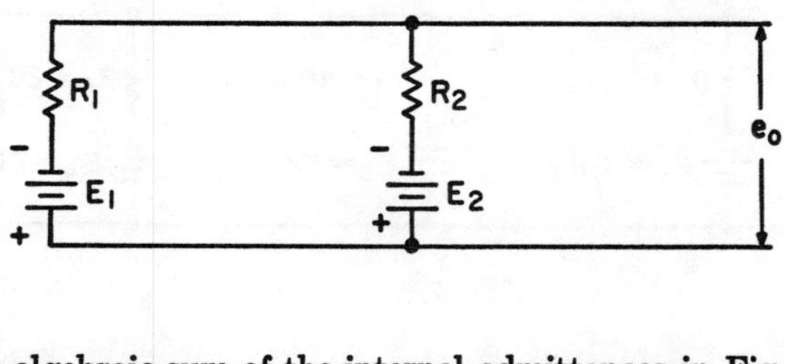

Fig. 7.1.

The algebraic sum of the internal admittances in Fig. 7.1 is equal to the reciprocal of R_1 plus the reciprocal of R_2. The algebraic sum of the short circuit currents in Fig. 7.1 is equal to E_1 divided by R_1 plus E_2 divided by R_2. Therefore,

$$e_o = \frac{E_1/R_1 + E_2/R_2}{1/R_1 + 1/R_2}$$

The student will find Millman's theorem a help in solving vacuum-tube circuits, three-phase circuits, and computer circuits.

Problems

7.1 Find the voltage across R_3 in Fig. 7.2.

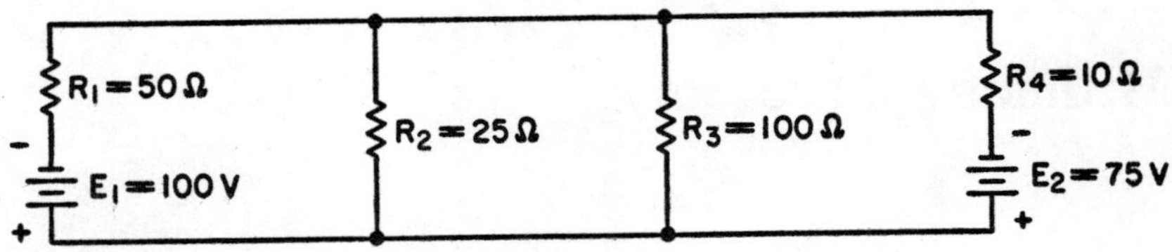

Fig. 7.2.

Answer

$$E_{R3} = \frac{\frac{E_1}{R_1} + \frac{E_2}{R_4}}{\frac{1}{R_1} + \frac{1}{R_2} + \frac{1}{R_3} + \frac{1}{R_4}} = \frac{\frac{100}{50} + \frac{75}{10}}{\frac{1}{50} + \frac{1}{25} + \frac{1}{100} + \frac{1}{10}}$$

$$E_{R3} = \frac{\frac{200 + 750}{100}}{\frac{2 + 4 + 1 + 10}{100}} = \frac{950}{17} = 55.9 \text{ volts}$$

7.2 Find e_o in Fig. 7.3.

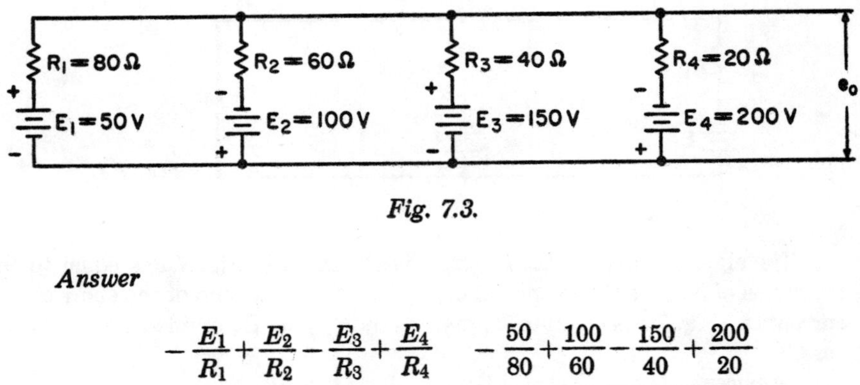

Fig. 7.3.

Answer

$$e_o = \frac{-\frac{E_1}{R_1} + \frac{E_2}{R_2} - \frac{E_3}{R_3} + \frac{E_4}{R_4}}{\frac{1}{R_1} + \frac{1}{R_2} + \frac{1}{R_3} + \frac{1}{R_4}} = \frac{-\frac{50}{80} + \frac{100}{60} - \frac{150}{40} + \frac{200}{20}}{\frac{1}{80} + \frac{1}{60} + \frac{1}{40} + \frac{1}{20}}$$

$$e_o = \frac{\frac{-150 + 400 - 900 + 2{,}400}{240}}{\frac{3 + 4 + 6 + 12}{240}} = \frac{1{,}750}{25} = 70 \text{ volts}$$

7.3 Find the current flowing through R_2 in Fig. 7.4.

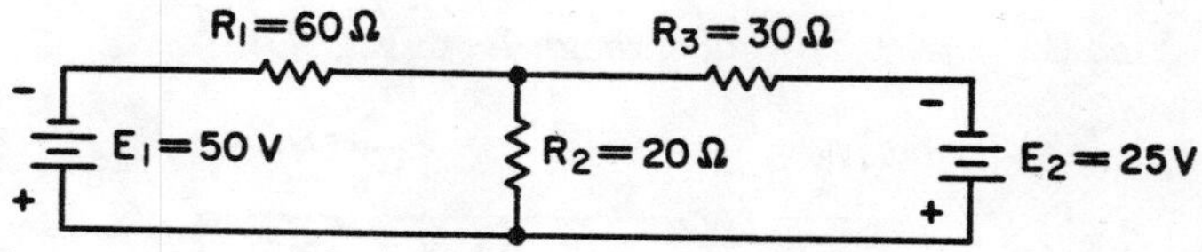

Fig. 7.4.

Answer

$$E_{R2}=\frac{\frac{E_1}{R_1}+\frac{E_2}{R_3}}{\frac{1}{R_1}+\frac{1}{R_2}+\frac{1}{R_3}}=\frac{\frac{50}{60}+\frac{25}{30}}{\frac{1}{60}+\frac{1}{20}+\frac{1}{30}}=\frac{\frac{50+50}{60}}{\frac{1+3+2}{60}}$$

$$E_{R2}=100/6=16.67 \text{ volts}$$

$$I_{R2}=E_{R_2}/R_2=16.67/20=0.834 \text{ amp}$$

7.4 Find the voltage e_o in Fig. 7.5.

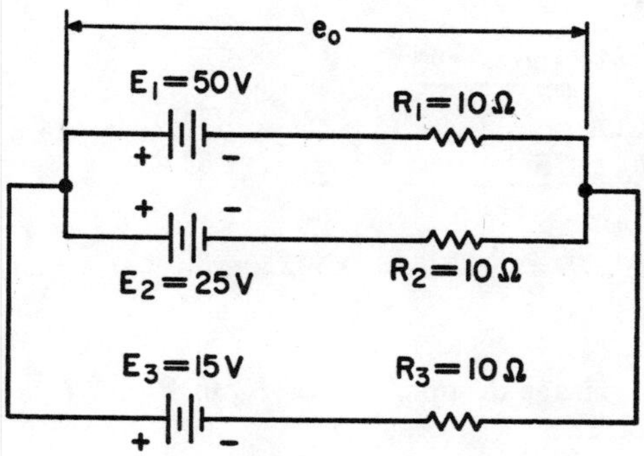

Fig. 7.5.

Answer

$$e_o=\frac{\frac{E_1}{R_1}+\frac{E_2}{R_2}+\frac{E_3}{R_3}}{\frac{1}{R_1}+\frac{1}{R_2}+\frac{1}{R_3}}=\frac{\frac{50}{10}+\frac{25}{10}+\frac{15}{10}}{\frac{1}{10}+\frac{1}{10}+\frac{1}{10}}$$

$$e_o=\frac{\frac{50+25+15}{10}}{\frac{1+1+1}{10}}=\frac{90}{3}=30 \text{ volts}$$

7.5 Find the current flowing through R_3 in Fig. 7.6.

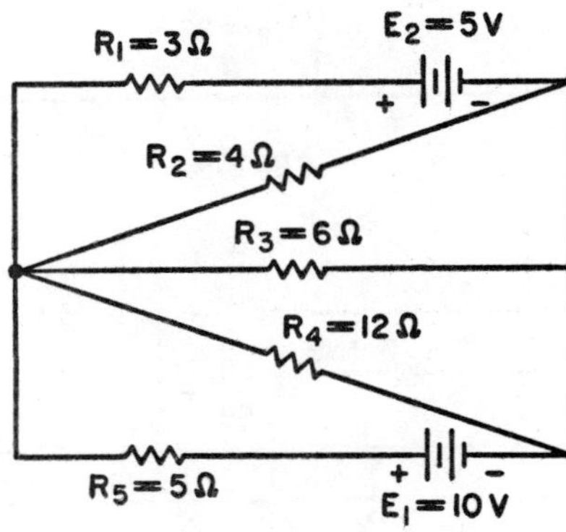

Fig. 7.6.

Answer

$$E_{R3} = \frac{\frac{E_2}{R_1} + \frac{E_1}{R_5}}{\frac{1}{R_1} + \frac{1}{R_2} + \frac{1}{R_3} + \frac{1}{R_4} + \frac{1}{R_5}} = \frac{\frac{5}{3} + \frac{10}{5}}{\frac{1}{3} + \frac{1}{4} + \frac{1}{6} + \frac{1}{12} + \frac{1}{5}}$$

$$E_{R3} = \frac{\frac{100 + 120}{60}}{\frac{20 + 15 + 10 + 5 + 12}{60}} = \frac{220}{62} = 3.55 \text{ volts}$$

$$I_{R3} = E_{R3}/R_3 = 3.55/6 = 0.592 \text{ amp}$$

7.6 Find the voltage dropped across R_2 in Fig. 7.7.

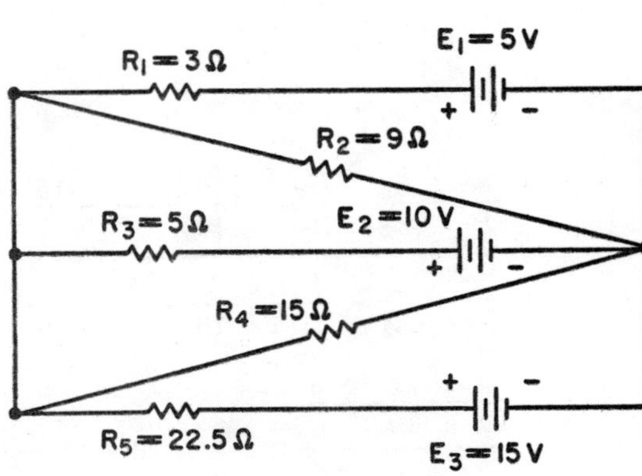

Fig. 7.7.

Answer

$$E_{R2} = \frac{\dfrac{E_1}{R_1} + \dfrac{E_2}{R_3} + \dfrac{E_3}{R_5}}{\dfrac{1}{R_1} + \dfrac{1}{R_2} + \dfrac{1}{R_3} + \dfrac{1}{R_4} + \dfrac{1}{R_5}} = \frac{\dfrac{5}{3} + \dfrac{10}{5} + \dfrac{15}{22.5}}{\dfrac{1}{3} + \dfrac{1}{9} + \dfrac{1}{5} + \dfrac{1}{15} + \dfrac{1}{22.5}}$$

$$E_{R2} = \frac{\dfrac{75 + 90 + 30}{45}}{\dfrac{15 + 5 + 9 + 3 + 2}{45}} = \frac{195}{34} = 5.74 \text{ volts}$$

7.7 Find e_o in Fig. 7.8.

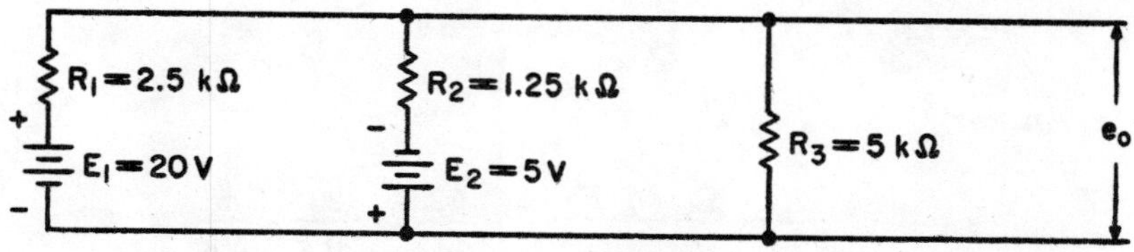

Fig. 7.8.

Answer

$$e_o = \frac{\dfrac{E_1}{R_1} - \dfrac{E_2}{R_2}}{\dfrac{1}{R_1} + \dfrac{1}{R_2} + \dfrac{1}{R_3}} = \frac{\dfrac{20}{2.5\text{ k}} - \dfrac{5}{1.25\text{ k}}}{\dfrac{1}{2.5\text{ k}} + \dfrac{1}{1.25\text{ k}} + \dfrac{1}{5\text{ k}}}$$

$$e_o = \frac{\dfrac{40 - 20}{5\text{ k}}}{\dfrac{2 + 4 + 1}{5\text{ k}}} = \frac{20}{7} = 2.86 \text{ volts}$$

7.8 If the switch is open, find e_o in Fig. 7.9.

Answer

$$e_o = \frac{\dfrac{E_1}{R_1} + \dfrac{E_2}{R_2}}{\dfrac{1}{R_1} + \dfrac{1}{R_2}} = \frac{\dfrac{120}{40\text{ k}} + \dfrac{20}{20\text{ k}}}{\dfrac{1}{40\text{ k}} + \dfrac{1}{20\text{ k}}} = \frac{\dfrac{120 + 40}{40\text{ k}}}{\dfrac{1 + 2}{40\text{ k}}}$$

$e_o = 160/3 = 53.3$ volts

7.9 If the switch is closed, find e_o in Fig. 7.9.

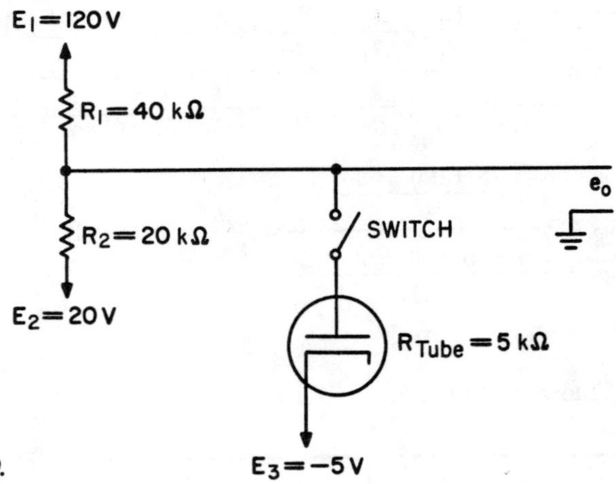

Fig. 7.9.

Answer

$$e_o = \frac{\frac{E_1}{R_1} + \frac{E_2}{R_2} - \frac{E_3}{R_{\text{Tube}}}}{\frac{1}{R_1} + \frac{1}{R_2} - \frac{1}{R_{\text{Tube}}}} = \frac{\frac{120}{40\text{ k}} + \frac{20}{20\text{ k}} - \frac{5}{5\text{ k}}}{\frac{1}{40\text{ k}} + \frac{1}{20\text{ k}} + \frac{1}{5\text{ k}}}$$

$$e_o = \frac{\frac{120 + 40 - 40}{40\text{ k}}}{\frac{1 + 2 + 8}{40\text{ k}}} = \frac{120}{11} = 10.9 \text{ volts}$$

7.10 Find e_o in Fig. 7.10.

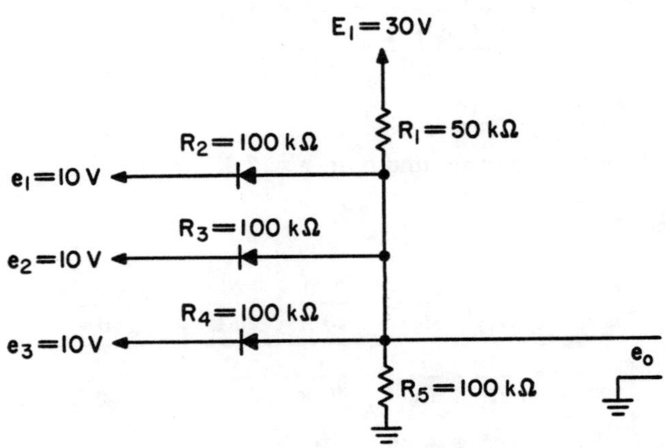

Fig. 7.10.

Answer

$$e_o = \frac{\dfrac{E_1}{R_1}+\dfrac{e_1}{R_2}+\dfrac{e_2}{R_3}+\dfrac{e_3}{R_4}}{\dfrac{1}{R_1}+\dfrac{1}{R_2}+\dfrac{1}{R_3}+\dfrac{1}{R_4}+\dfrac{1}{R_5}} = \frac{\dfrac{30}{50\text{ k}}+\dfrac{10}{100\text{ k}}+\dfrac{10}{100\text{ k}}+\dfrac{10}{100\text{ k}}}{\dfrac{1}{50\text{ k}}+\dfrac{1}{100\text{ k}}+\dfrac{1}{100\text{ k}}+\dfrac{1}{100\text{ k}}+\dfrac{1}{100\text{ k}}}$$

$$e_o = \frac{\dfrac{60+10+10+10}{100\text{ k}}}{\dfrac{2+1+1+1+1}{100\text{ k}}} = \frac{90}{6} = 15\text{ volts}$$

7.11 Find e_o in Fig. 7.11.

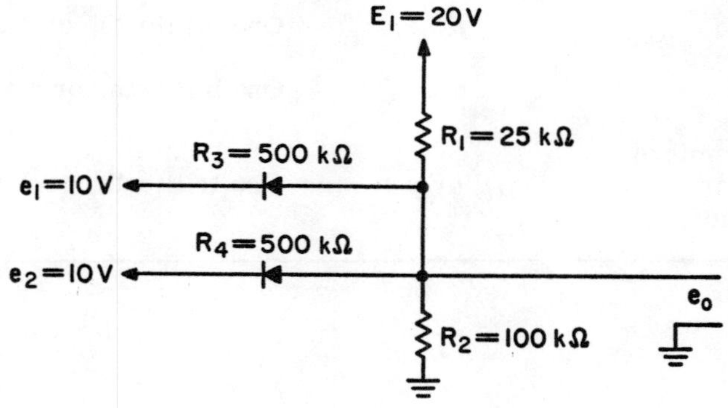

Fig. 7.11.

Answer

$$e_o = \frac{\dfrac{E_1}{R_1}+\dfrac{e_1}{R_3}+\dfrac{e_2}{R_4}}{\dfrac{1}{R_1}+\dfrac{1}{R_2}+\dfrac{1}{R_3}+\dfrac{1}{R_4}} = \frac{\dfrac{20}{25\text{ k}}+\dfrac{10}{500\text{ k}}+\dfrac{10}{500\text{ k}}}{\dfrac{1}{25\text{ k}}+\dfrac{1}{100\text{ k}}+\dfrac{1}{500\text{ k}}+\dfrac{1}{500\text{ k}}}$$

$$e_o = \frac{\dfrac{400+10+10}{500\text{ k}}}{\dfrac{20+5+1+1}{500\text{ k}}} = \frac{420}{27} = 15.55\text{ volts}$$

Appendix A

Conversion Factors

Prefix	Symbol	Unit
Giga	G	One billion, or 1×10^{9}
Mega	M	One million, or 1×10^{6}
Kilo	K	One thousand, or 1×10^{3}
milli	m	One thousandth, or 1×10^{-3}
micro	μ	One millionth, or 1×10^{-6}
nano	n	One billionth, or 1×10^{-9}
micromicro or pico	$\mu\mu$ or p	One trillionth, or 1×10^{-12}

Appendix B
Trigonometrical Ratios for Angles from 0° to 90°

Tangent	Angle	Sine	Cosine		
0.0000	0°	0.	1.0000	90°	∞
.0175	1°	.0175	.9998	89°	57.29
.0349	2°	.0349	.9994	88°	28.6363
.0524	3°	.0523	.9986	87°	19.0811
.0699	4°	.0698	.9976	86°	14.3007
.0875	5°	.0872	.9962	85°	11.4301
.1051	6°	.1045	.9945	84°	9.5144
.1228	7°	.1219	.9925	83°	8.1443
.1405	8°	.1392	.9903	82°	7.1154
.1584	9°	.1564	.9877	81°	6.3138
.1763	10°	.1736	.9848	80°	5.6713
.1944	11°	.1908	.9816	79°	5.1446
.2126	12°	.2079	.9781	78°	4.7046
.2309	13°	.2250	.9744	77°	4.3315
.2493	14°	.2419	.9703	76°	4.0108
.2679	15°	.2588	.9659	75°	3.7321
.2867	16°	.2756	.9613	74°	3.4874
.3057	17°	.2924	.8563	73°	3.2709
.3249	18°	.3090	.9511	72°	3.0777
.3443	19°	.3256	.9455	71°	2.9042
.3640	20°	.3420	.9397	70°	2.7475
.3839	21°	.3584	.9336	69°	2.6051
.4040	22°	.3746	.9272	68°	2.4751
.4245	23°	.3907	.9205	67°	2.3559
.4452	24°	.4067	.9135	66°	2.2460
.4663	25°	.4226	.9063	65°	2.1445
.4877	26°	.4384	.8988	64°	2.0503
.5095	27°	.4540	.8910	63°	1.9626
.5317	28°	.4695	.8829	62°	1.8807
.5543	29°	.4848	.8746	61°	1.8040
.5774	30°	.5000	.8660	60°	1.7321
.6009	31°	.5150	.8572	59°	1.6643
.6249	32°	.5299	.8480	58°	1.6003
.6494	33°	.5446	.8387	57°	1.5399
.6745	34°	.5592	.8290	56°	1.4826
.7002	35°	.5736	.8192	55°	1.4281
.7265	36°	.5878	.8090	54°	1.3764
.7536	37°	.6018	.7986	53°	1.3270
.7813	38°	.6157	.7880	52°	1.2799
.8098	39°	.6293	.7771	51°	1.2349
.8391	40°	.6428	.7660	50°	1.1918
.8693	41°	.6561	.7547	49°	1.1504
.9004	42°	.6691	.7431	48°	1.1106
.9325	43°	.6820	.7314	47°	1.0724
.9657	44°	.6947	.7193	46°	1.0355
1.000	45°	.7071	.7071	45°	1.0000
		Cosine	Sine	Angle	Tangent

Appendix C

Four-Figure Logarithms

	0	1	2	3	4	5	6	7	8	9	1	2	3	4	5	6	7	8	9
10	0000	0043	0086	0128	0170	0212	0253	0294	0334	0374	4 4	9 8	13 12	17 16	21 20	26 24	30 28	34 32	38 36
11	0414	0453	0492	0531	0569	0607	0645	0682	0719	0755	4 4	8 7	12 11	15 15	19 19	23 22	27 26	31 30	35 33
12	0792	0828	0864	0899	0934	0969	1004	1038	1072	1106	3 3	7 7	11 10	14 14	18 17	21 20	25 24	28 27	32 31
13	1139	1173	1206	1239	1271	1303	1335	1367	1399	1430	3 3	7 7	10 10	13 13	16 16	20 19	23 22	26 25	30 29
14	1461	1492	1523	1553	1584	1614	1644	1673	1703	1732	3 3	6 6	9 9	12 12	15 15	19 17	22 20	25 23	28 26
15	1761	1790	1818	1847	1875	1903	1931	1959	1987	2014	3 3	6 6	9 8	11 11	14 14	17 17	20 19	23 22	26 25
16	2041	2068	2095	2122	2148	2175	2201	2227	2253	2279	3 3	5 5	8 8	11 10	14 13	16 16	19 18	22 21	24 23
17	2304	2330	2355	2380	2405	2430	2455	2480	2504	2529	3 2	5 5	8 7	10 10	13 12	15 15	18 17	20 20	23 22
18	2553	2577	2601	2625	2648	2672	2695	2718	2742	2765	2 2	5 5	7 7	9 9	12 11	14 14	16 16	19 18	21 21
19	2788	2810	2833	2856	2878	2900	2923	2945	2967	2989	2 2	4 4	7 6	9 8	11 11	13 13	16 15	18 17	20 19
20	3010	3032	3054	3075	3096	3118	3139	3160	3181	3201	2	4	6	8	11	13	15	17	19
21	3222	3243	3263	3284	3304	3324	3345	3365	3385	3404	2	4	6	8	10	12	14	16	18
22	3424	3444	3464	3483	3502	3522	3541	3560	3579	3598	2	4	6	8	10	12	14	15	17
23	3617	3636	3655	3674	3692	3711	3729	3747	3766	3784	2	4	6	7	9	11	13	15	17
24	3802	3820	3838	3856	3874	3892	3909	3927	3945	3962	2	4	5	7	9	11	12	14	16
25	3979	3997	4014	4031	4048	4065	4082	4099	4116	4133	2	3	5	7	9	10	12	14	51
26	4150	4166	4183	4200	4216	4232	4249	4265	4281	4298	2	3	5	7	8	10	11	13	15
27	4314	4330	4346	4362	4378	4393	4409	4425	4440	4456	2	3	5	6	8	9	11	13	14
28	4472	4487	4502	4518	4533	4548	4564	4579	4594	4609	2	3	5	6	8	9	11	12	14
29	4624	4639	4654	4669	4683	4698	4713	4728	4742	4757	1	3	4	6	7	9	10	12	13
30	4771	4786	4800	4814	4829	4843	4857	4871	4886	4900	1	3	4	6	7	9	10	11	13
31	4914	4928	4942	4953	4969	4983	4997	5011	5024	5038	1	3	4	6	7	8	10	11	12
32	5051	5065	5079	5092	5105	5119	5132	5145	5159	5172	1	3	4	5	7	8	9	11	12
33	5185	5198	5211	5224	5237	5250	5263	5276	5289	5302	1	3	4	5	6	8	9	10	12
34	5315	5328	5340	5353	5366	5378	5391	5403	5416	5428	1	3	4	5	6	8	9	10	11
35	5441	5453	5465	5478	5490	5502	5514	5527	5539	5551	1	2	4	5	6	7	9	10	11
36	5563	5575	5587	5599	5611	5623	5635	5647	5658	5670	1	2	4	5	6	7	8	10	11
37	5682	5694	5705	5717	5729	5740	5752	5763	5775	5786	1	2	3	5	6	7	8	9	10
38	5798	5809	5821	5832	5843	5855	5866	5877	5888	5899	1	2	3	5	6	7	8	9	10
39	5911	5922	5933	5944	5955	5966	5977	5988	5999	6010	1	2	3	4	5	7	8	9	10
40	6021	6031	6042	6053	6064	6075	6085	6096	6107	6117	1	2	3	4	5	6	8	9	10
41	6128	6138	6149	6160	6170	6180	6191	6201	6212	6222	1	2	3	4	5	6	7	8	9
42	6232	6243	6253	6263	6274	6284	6294	6304	6314	6325	1	2	3	4	5	6	7	8	9
43	6335	6345	6355	6365	6375	6385	6395	6405	6415	6425	1	2	3	4	5	6	7	8	9
44	6435	6444	6454	6464	6474	6484	6493	6503	6513	6522	1	2	3	4	5	6	7	8	9
45	6532	6542	6551	6561	6571	6580	6590	6599	6609	6618	1	2	3	4	5	6	7	8	9
46	6628	6637	6646	6656	6665	6675	6684	6693	6702	6712	1	2	3	4	5	6	7	7	8
47	6721	6730	6739	6749	6758	6767	6776	6785	6794	6803	1	2	3	4	5	5	6	7	8
48	6812	6821	6830	6839	6848	6857	6866	6875	6884	6893	1	2	3	4	4	5	6	7	8
49	6902	6911	6920	6928	6937	6946	6955	6964	6972	6981	1	2	3	4	4	5	6	7	8

Four-Figure Logarithms (cont'd)

	0	1	2	3	4	5	6	7	8	9	1	2	3	4	5	6	7	8	9
50	6990	6998	7007	7016	7024	7033	7042	7050	7059	7067	1	2	3	3	4	5	6	7	8
51	7076	7084	7093	7101	7110	7118	7126	7135	7143	7152	1	2	3	3	4	5	6	7	8
52	7160	7168	7177	7185	7193	7202	7210	7218	7226	7235	1	2	2	3	4	5	6	7	7
53	7243	7251	7259	7267	7275	7284	7292	7300	7308	7316	1	2	2	3	4	5	6	6	7
54	7324	7332	7340	7348	7356	7364	7372	7380	7388	7396	1	2	2	3	4	5	6	6	7
55	7404	7412	7419	7427	7435	7443	7451	7459	7466	7474	1	2	2	3	4	5	5	6	7
56	7482	7490	7497	7505	7513	7520	7528	7536	7543	7551	1	2	2	3	4	5	5	6	7
57	7559	7566	7574	7582	7589	7597	7604	7612	7619	7627	1	2	2	3	4	5	5	6	7
58	7634	7642	7649	7657	7664	7672	7679	7686	7694	7701	1	1	2	3	4	4	5	6	7
59	7709	7716	7723	7731	7738	7745	7752	7760	7767	7774	1	1	2	3	4	4	5	6	7
60	7782	7789	7796	7803	7810	7818	7825	7832	7839	7846	1	1	2	3	4	4	5	6	6
61	7853	7860	7868	7875	7882	7889	7896	7903	7910	7917	1	1	2	3	4	4	5	6	6
62	7924	7931	7938	7945	7952	7959	7966	7973	7980	7987	1	1	2	3	3	4	5	6	6
63	7993	8000	8007	8014	8021	8028	8035	8041	8048	8055	1	1	2	3	3	4	5	5	6
64	8062	8069	8075	8082	8089	8096	8102	8109	8116	8122	1	1	2	3	3	4	5	5	6
65	8129	8136	8142	8149	8156	8162	8169	8176	8182	8189	1	1	2	3	3	4	5	5	6
66	8195	8202	8209	8215	8222	8228	8235	8241	8248	8254	1	1	2	3	3	4	5	5	6
67	8261	8267	8274	8280	8287	8293	8299	8306	8312	8319	1	1	2	3	3	4	5	5	6
68	8325	8331	8338	8344	8351	8357	8363	8370	8376	8382	1	1	2	3	3	4	4	5	6
69	8388	8395	8401	8407	8414	8420	8426	8432	8439	8445	1	1	2	2	3	4	4	5	6
70	8451	8457	8463	8470	8476	8482	8488	8494	8500	8506	1	1	2	2	3	4	4	5	6
71	8513	8519	8525	8531	8537	8543	8549	8555	8561	8567	1	1	2	2	3	4	4	5	5
72	8573	8579	8585	8591	8597	8603	8609	8615	8621	8627	1	1	2	2	3	4	4	5	5
73	8633	8639	8645	8651	8657	8663	8669	8675	8681	8686	1	1	2	2	3	4	4	5	5
74	8692	8698	8704	8710	8716	8722	8727	8733	8739	8745	1	1	2	2	3	4	4	5	5
75	8751	8756	8762	8768	8774	8779	8785	8791	8797	8802	1	1	2	2	3	3	4	5	5
76	8808	8814	8820	8825	8831	8837	8842	8848	8854	8859	1	1	2	2	3	3	4	5	5
77	8865	8871	8876	8882	8887	8893	8899	8904	8910	8915	1	1	2	2	3	3	4	4	5
78	8921	8927	8932	8938	8943	8949	8954	8960	8965	8971	1	1	2	2	3	3	4	4	5
79	8976	8982	8987	8993	8998	9004	9009	9015	9020	9025	1	1	2	2	3	3	4	4	5
80	9031	9036	9042	9047	9053	9058	9063	9069	9074	9079	1	1	2	2	3	3	4	4	5
81	9085	9090	9096	9101	9106	9112	9117	9122	9128	9133	1	1	2	2	3	3	4	4	5
82	9138	9143	9149	9154	9159	9165	9170	9175	9180	9186	1	1	2	2	3	3	4	4	5
83	9191	9196	9201	9206	9212	9217	9222	9227	9232	9238	1	1	2	2	3	3	4	4	5
84	9243	9248	9253	9258	9263	9269	9274	9279	9284	9289	1	1	2	2	3	3	4	4	5
85	9294	9299	9304	9309	9315	9320	9325	9330	9335	9340	1	1	2	2	3	3	4	4	5
86	9345	9350	9355	9360	9365	9370	9375	9380	9385	9390	1	1	2	2	3	3	4	4	5
87	9395	9400	9405	9410	9415	9420	9425	9430	9435	9440	0	1	1	2	2	3	3	4	4
88	9445	9450	9455	9460	9465	9469	9474	9479	9484	9489	0	1	1	2	2	3	3	4	4
89	9494	9499	9504	9509	9513	9518	9523	9528	9533	9538	0	1	1	2	2	3	3	4	4
90	9542	9547	9552	9557	9562	9566	9571	9576	9581	9586	0	1	1	2	2	3	3	4	4
91	9590	9595	9600	9605	9609	9614	9619	9624	9628	9633	0	1	1	2	2	3	3	4	4
92	9638	9643	9647	9652	9657	9661	9666	9671	9675	9680	0	1	1	2	2	3	3	4	4
93	9685	9689	9694	9699	9703	9708	9713	9717	9722	9727	0	1	1	2	2	3	3	4	4
94	9731	9736	9741	9745	9750	9754	9759	9763	9768	9773	0	1	1	2	2	3	3	4	4
95	9777	9782	9786	9791	9795	9800	9805	9809	9814	9818	0	1	1	2	2	3	3	4	4
96	9823	9827	9832	9836	9841	9845	9850	9854	9859	9863	0	1	1	2	2	3	3	4	4
97	9868	9872	9877	9881	9886	9890	9894	9899	9903	9908	0	1	1	2	2	3	3	4	4
98	9912	9917	9921	9926	9930	9934	9939	9943	9948	9952	0	1	1	2	2	3	3	4	4
99	9956	9961	9965	9969	9974	9978	9983	9987	9991	9996	0	1	1	2	2	3	3	3	4

INDEX